AF737104

Neuroscience and Society

Advances in neuroscience research are rapidly redefining what it means to be human. The absence of the brain/mind dichotomy has, in turn, removed the separation between our brain biology and our sociocultural experiences, raising questions for social sciences to address. How responsible are we, as individuals, for our actions? Do we have free will? Is it ethical for us to peer into others' brains? How are our collective social cultural norms influenced by our brain function? At the same time, neuroscientists need to develop better intuition about the ethical, legal, and social implications of their research. Close collaboration between neuroscience and social sciences is the best way forward.

This book acts as an introduction to these and other issues that lie at the interface of neuroscience and social sciences, using the physiological underpinnings of our decision-making processes as a framework. Examples of topics addressed here are:

- Neuroscience and economics
- Neuroscience and law
- Neuroscience and ethics
- Neuroscience and mental health
- Neuroscience of religion and humour

The book is intended for students of neuroscience and social sciences, as well as readers generally interested in the human condition. It is hoped that the book will stimulate cross-disciplinary thinking and inspire a new generation of thinkers who are willing to look at both social sciences and neuroscience research with a different lens. Such bridge builders will be the pioneers of the next level of interrogation at this emerging interface.

Dr. Sukumar Vijayaraghavan is a neuroscientist and professor at the Department of Physiology and Biophysics, University of Colorado, School of Medicine. He has wide-ranging interests from synaptic transmission, olfaction, and drug addiction to graduate education and the interaction between neuroscience and social sciences.

Dr. Gidon Felsen is a professor in the Department of Physiology and Biophysics at the University of Colorado School of Medicine. His research focuses on the neural mechanisms of decisions and actions under normal and pathological conditions and on how neuroscience can inform societally relevant questions.

Foundations and Innovations in Neurobiology

Formerly *Frontiers in Neuroscience*

Series Editor: Lynne M. Bianchi

Neurobiology remains one of the most intriguing areas of research, and scientists and clinicians from an array of backgrounds and subspecialties strive to understand everything from the intricate molecular pathways governing activity of a single cell to the complex connections that underlie aspects of human behavior.

As new discoveries continually shape our understanding of how the nervous system forms and functions, neuroscientists depend on reliable resources to summarize the recent progress. The purpose of this reference-oriented series is to capture some of the most important advances in neurobiology by showcasing groundbreaking discoveries, current theories, and future directions. Each volume focuses on a core area of neurobiology and features contributions from experts in the subfield discussed. Whether the book is devoted to a cellular function, brain region, clinical challenge, or new methodology, each volume provides the reader with a strong foundation and deeper appreciation for the current state of the field.

Neuroscience and Society
The Interface of Neuroscience and the Social Sciences

Edited by
Sukumar Vijayaraghavan and Gidon Felsen

CRC Press
Taylor & Francis Group
Boca Raton London New York

CRC Press is an imprint of the
Taylor & Francis Group, an **informa** business

Designed cover image: Getty Images

First edition published 2026
by CRC Press
2385 NW Executive Center Drive, Suite 320, Boca Raton FL 33431

and by CRC Press
4 Park Square, Milton Park, Abingdon, Oxon, OX14 4RN

CRC Press is an imprint of Taylor & Francis Group, LLC

© 2026 Taylor & Francis Group, LLC

© 2026 selection and editorial matter, Sukumar Vijayaraghavan and Gidon Felsen; individual chapters, the contributors

Reasonable efforts have been made to publish reliable data and information, but the author and publisher cannot assume responsibility for the validity of all materials or the consequences of their use. The authors and publishers have attempted to trace the copyright holders of all material reproduced in this publication and apologize to copyright holders if permission to publish in this form has not been obtained. If any copyright material has not been acknowledged please write and let us know so we may rectify in any future reprint.

Except as permitted under U.S. Copyright Law, no part of this book may be reprinted, reproduced, transmitted, or utilized in any form by any electronic, mechanical, or other means, now known or hereafter invented, including photocopying, microfilming, and recording, or in any information storage or retrieval system, without written permission from the publishers.

For permission to photocopy or use material electronically from this work, access www.copyright.com or contact the Copyright Clearance Center, Inc. (CCC), 222 Rosewood Drive, Danvers, MA 01923, 978–750–8400. For works that are not available on CCC please contact mpkbookspermissions@tandf.co.uk

Trademark notice: Product or corporate names may be trademarks or registered trademarks and are used only for identification and explanation without intent to infringe.

ISBN: 978-1-032-70917-8 (hbk)
ISBN: 978-1-032-71778-4 (pbk)
ISBN: 978-1-032-71779-1 (ebk)

DOI: 10.1201/9781032717791

Typeset in Corbel
by Apex CoVantage, LLC

Contents

Contributors

Ted Doykos
Department of Physiology and Biophysics
University of Colorado School of Medicine
Aurora, Colorado

Kathinka Evers
Center for Research Ethics and Bioethics
Uppsala University
Uppsala, Sweden

Michele Farisco
Center for Research Ethics and Bioethics
Uppsala University
Uppsala, Sweden

Gidon Felsen
Department of Physiology and Biophysics
University of Colorado School of Medicine
Aurora, Colorado

Guido Frank
Department of Psychiatry
University of California San Diego
San Diego, California

Shelia Kennison
Department of Psychology
Oklahoma State University
Stillwater, Oklahoma

Karolina Lempert
Gordon F. Dernier School of Psychology
Adelphi University
Garden City, New York

Gerben Meynen
Department of Philosophy
Vrije Universiteit Amsterdam
Amsterdam, Netherlands

John Monterosso
Department of Psychology
University of Southern California
Los Angeles, California

Emily R.D. Murphy
College of Law
University of California
San Francisco, California

Andrew Newberg
Department of Integrative Medicine and Nutritional Sciences
Thomas Jefferson University
Philadelphia, Pennsylvania

Natalia Peraza
Department of Psychology
University of Southern California
Los Angeles, California

Joel Stoddard
Department of Psychiatry
University of Colorado School of Medicine
Aurora, Colorado

Sukumar Vijayaraghavan
Department of Physiology and Biophysics
University of Colorado School of Medicine
Aurora, Colorado

Preface

Once we dispense with the mind–brain dichotomy, a society is just a grouping of brains brought together by geography, interests, culture, and history. Thus, it comes as no surprise that neuroscience research has implications for all of society, providing a new way to look at social issues. In this book, we bring a collection of chapters that provide an overview of the intersection between neuroscience and social sciences. While not extensive, the book provides a synopsis of a number of issues raised by progress in neuroscience that acts as a primer to interested students and members of the public.

Social sciences have had an uncomfortable relationship with neuroscience. The apprehension is that the reduction of humans to brain biology strips away all the social, cultural, and interpersonal aspects that enrich human experience. This is due to a misunderstanding of what neuroscience does. As stated by Fitzgerald and Callard (2015), it is not tenable to maintain a separation of sociocultural and neurobiological aspects of our existence. The optimal solution is for social scientists to incorporate lessons from neuroscience and its approaches in their epistemological endeavours. Indeed, an individual's worldview is entirely a product of his/her brain function based on inputs from the environment and society at large. The brain takes these inputs and alters its own function within the set points provided by genetics, epigenetics, and prior experiences. This synthesis contributes to our cultural, ethical, and social norms that are the targets of exploration in the social sciences. The intricate association between our individual experiences and the social gestalt cannot be overstated, nor can the two be delinked.

If social sciences need neuroscience, the reverse is true as well. In 1949, the Nobel Prize for Physiology and Medicine was awarded to the Portuguese neurologist Antonio Egas Moniz for the discovery and use of prefrontal lobotomy for grossly disturbed psychiatric patients. The purpose was not to remedy the illness or restore sanity but to calm the patients. Thousands of these procedures were performed between 1940 and early 1960s. The procedure was fraught with ethical concerns, with many patients suffering from drastic cognitive and personality changes. In 1950, the Soviet Union was the first country to ban the procedure, considering it contrary to the principles of humanity. The opposition brought to bear ethics, Stalinist politics, and sociology, as well as personal and professional rivalries, in their successful bid to outlaw the procedure (Zajicek, 2017). This interplay of social, cultural, and scientific arguments is constantly played out even today. For example, there is general leeriness among people when it comes to human manipulations. A few decades ago, the fear revolved around recombinant DNA research, leading the bioethicist Willard Gaylin to coin the term "the Frankenstein Factor" (1977). This is now more so with neurosciences where the spectre of behavior control and brain modifications sounds ominous (Gaylin, 2009). The onus is on neuroscientists and social scientists working in concert to draft policies and protocols to allay these fears. In essence, the neuroscientist answers the question of how we should do something, while the social scientist helps with the question of whether we should do it.

The field of neuroscience and society explores the idea that advancement in the field should take us in the right direction that is for the good of society. It is the thinking that science should align with society's values; thus, the term 'good for the public' is determined by the public itself (Kaebnick, 2024). Thus, the public needs to be engaged in both prophetic as well as regulatory endeavours. The importance of public engagement cannot be overemphasized, and neuroscientists need to be educated in the timely and accurate dissemination of information to the public in an accessible and timely manner.

Another issue that needs better communication is the time it takes for a discovery in science, neuroscience included, to actually result in implementation. In 2011, Stephen Morse, Professor of Law and Psychology wrote

> The genome was fully sequenced in 2001, and there has not been one resulting major advance in therapeutic medicine since.
>
> **(Morse, 2011)**

However, we have made tremendous advances in precision and personalized medicine since 2001 based on the human genome. For example, comprehensive genomic profiling has resulted in much more effective strategies in individualized cancer treatment (Pankiw et al., 2024). Conversely, while neuroscience research definitively showed the biological bases for drug addiction by the end of 1970s, drug offences are still treated as a personal and moral failing by our legal system. There is, thus, a need for more effective communication between neuroscience and social sciences, in parallel with public education, to implement lessons learnt from our studies on brain function to society as a whole.

This book raises some of these issues that stand at the interface of neuroscience and society. The purpose of this book is to make students of neuroscience and social sciences aware of this interface as areas for investigation as they progress in their careers. The aim is to nurture a cadre of bridge builders who have the ability to move out of their respective academic silos. Towards that end, the book addresses issues in economics, law, ethics, and other areas that are fertile grounds for such interdisciplinary endeavours.

References

Fitzgerald, D., and Callard, F. (2015). Social science and neuroscience beyond interdisciplinarity: Experimental entanglements. *Theory Cult. Soc.*, 32, 3–32.

Gaylin, W. (1977). The Frankenstein factor. *New Eng. J. Med.*, 297, 665–667.

Gaylin, W. (2009). Behavior control: From the brain to the mind. *Hastings Cent. Rep.*, 39, 13–16.

Kaebnick, G.E. (2024). Neuroscience and society: Supporting and unsettling public engagement. *Hastings Cent. Rep.*, 54, 20–23.

Morse, S.J. (2011). Avoiding irrational neurolaw exuberance: A plea for neuromodesty. *Mercer Law Rev.*, 62 (3), 837–859.

Pankiw, M., Brezden-Masley, C., and Charames, G.S. (2024). Comprehensive genomic profiling for oncological advancements by precision medicine. *Med. Oncol.*, 41, 1–13.

Zajicek, B. (2017). Banning the Soviet Lobotomy: Psychiatry, ethics, and professional politics during late Stalinism. *Bull. Hist. Med.*, 91, 33–61.

1

Introduction: The Neuroscience of Decision Making

Ted Doykos and Gidon Felsen

> Albus Dumbledore: "It is our choices . . . that show what we truly are, far more than our abilities."
>
> **– J. K. Rowling, *Harry Potter and the Chamber of Secrets***
>
> Alice: "Would you tell me, please, which way I ought to go from here?"
>
> Cheshire Cat: "That depends a good deal on where you want to get to."
>
> **– Lewis Carroll, *Alice's Adventures in Wonderland***

Humans are innately social creatures. We have learned to operate within the shared norms of increasingly complex social groups, from the small bands of our ancestors to our modern virtual communities. These social groups influence how our brains function, and brain functions are influenced by the environmental contexts in which they are embedded. In particular, the human brain evolved to promote survival during a period of acute vulnerability to natural disaster, disease, predation, and food scarcity, resulting in a cognitive bias towards considering nearby and imminent factors over distant and future factors (Greene, 2003). The means by which humans coped with these life-threatening challenges, through incremental advances in tool use and language (Donald, 1991), drove the innovative activities that culminated in the formation of the societies we live in today.

While modern society arose to address urgent human concerns, the rapid pace of technological development has replaced these concerns with "first-world problems" faster than our brains have evolved. Thus, we sometimes find ourselves interacting in a disorderly or uncivilized manner that would have made sense in the world of early humans but is counterproductive in modern society. For example, judging another as friend or foe based on physical traits was likely adaptive when early humans lived in homogeneous groups but runs counter to the values of pluralism in a multicultural society. Thus, despite the innumerable achievements made by modern society, the fact that our brains are adapted to a pre-modern world has ensured that conflict and suboptimal decision-making remain ubiquitous aspects of our experience. This dynamic interplay between brain function and various aspects of modern society constitutes the major theme of this book.

DOI: 10.1201/9781032717791-1

The book covers a wide range of topics, from humor (Chapter 10) to psychiatry (Chapters 5, 7) to artificial intelligence (Chapter 9). Each chapter can be read on its own, and the reader is free to skip to those topics most interesting to them. However, the purpose of this introductory chapter is to provide an overarching framework to bridge the gap between neuroscience and the specific societal topics of subsequent chapters. We propose that the neuroscience of decision making can provide such an overarching framework.

Why decision making? As concisely stated in our first epigraph, decision making is central to who we are and how we operate in society. At first glance, decision making may conjure images of particularly momentous choices ("What career should I pursue?"). However, when considered more broadly, we are constantly making decisions large and small, often below the level of conscious awareness ("Should I start walking with my left foot or my right?"). Across this spectrum, decision making, like all other brain functions, is subserved by basic physiological processes constrained by the chemical and physical laws of nature.

Modern neuroscience seeks to elucidate fundamental brain function at the level of basic physiological processes. The primary processing units in the nervous system are neurons, specialized cells that use electrical and chemical signals to communicate with one another. The human brain contains approximately 100 billion neurons, each of which communicates with many other neurons, forming neural circuits that subserve specific functions like representing sensory information and enacting motor plans. Electrical signals known as action potentials are the fundamental unit for representing and transmitting information in the nervous system (Bear et al., 2016). The rate at which any given neuron produces action potentials – "neural activity" – is driven by the activity of upstream neurons and varies with sensory stimuli, behavioral state, and other internal factors. Modern psychiatry increasingly seeks to understand disorders of the nervous system at a similar level of basic physiological processes, towards the goal of using tools from neuroscience for therapeutic treatment (see Chapter 7, "Neuroscience in Psychiatry").

Neural activity forms the basis for human decision making. According to one influential model, the activity exhibited by different groups of neurons represents competing choice options, with the activity in each group influenced by current sensory information, prior experience, and the subjective value of each option (Gold & Shadlen, 2007). In this model, making a particular decision corresponds to the activity in one of the groups crossing a critical threshold (Smith & Ratcliff, 2004). Groups representing competing choice options – for example, moving left or right – often actively inhibit one another, "competing" to cross their respective thresholds first. If the "move left" neurons win the competition, they drive the activity of downstream groups of neurons that ultimately initiate a leftward movement.

A wealth of evidence, primarily from animal model experiments in which neural activity is recorded and perturbed during simple decision-making tasks, supports this idea. In a series of seminal studies, activity in a region of visual cortex was examined in primates presented with a visual stimulus consisting of a cloud of dots, some of which were moving leftward and some rightward. The primate's task was to decide in which direction the majority of dots were moving and to report its decision with an eye movement in the corresponding direction (left or right). In one study, activity recorded from individual neurons as the visual stimulus was presented was found to correlate with the direction of stimulus motion and choices during the task: in fact, one could "predict", based on the activity of a single neuron, which choice (left vs. right) was made on each trial (Britten et al., 1992; Newsome et al., 1989). Subsequent studies found that electrically stimulating neural activity in the same brain area predictably altered the primate's choices, suggesting that activity in this region is causally related to, and not simply correlated with, decisions (Salzman et al., 1992). These and many other studies inspired an active and growing field examining the neural bases of decision making, with modern studies often recording and perturbing the activity of large populations of specific types of neurons across multiple brain regions during decision making tasks (Luo et al., 2018; Steinmetz et al., 2019).

Translating results from animals making simple sensory decisions in a laboratory environment to humans making societally relevant decisions in the wild is challenging. Methods examining human brain function have less spatiotemporal resolution compared to animal models, with rare exceptions afforded by neurosurgical procedures in which the brain is made accessible for clinical purposes (Herd et al., 2021; Tekriwal et al., 2019). Nevertheless, given that the basic structure and function of subcortical brain regions are evolutionarily conserved across mammalian species, it is likely that the mechanisms underlying simple decisions are similar between humans and animals. Further, while the mechanisms of higher-level decisions are still an active area of study, they likely evolved from low-level decisions and leverage the computational power of expanded cortical tissue available to humans (Okazawa & Kiani, 2023; Shadlen & Kiani, 2013).

Returning to our simple example decision, whether activity is higher in "move left" or "move right" neurons depends on several factors, including "bottom-up" inputs providing sensory information (e.g., "there is a tiger to my right") and "top-down" inputs representing behavioral goals (e.g., "those berries I was planning to eat are *also* to my right"). This example illustrates two related features of decision making that will prove useful in linking neuroscience with the societal questions examined in this book: Decision making is hierarchical and arises from the integration of top-down and bottom-up processes.

1.1 Decision Making Is Hierarchical

There is strong neuroscientific evidence that decision making is organized hierarchically (Miller & Cohen, 2001). Lower-level decisions are often made without conscious awareness and roughly map to activity in subcortical brain circuits. For example, circuits in the midbrain and brainstem direct attention based on incoming sensory stimuli. Conscious decision making is often subserved by higher-level cortical regions exerting top-down control over subcortical circuits to align choices with behavioral goals. Thus, different decisions can occur even when presented with identical sensory information. For example, whether you decide to speed up or slow down in response to a yellow traffic light may depend on whether you're running late. Both choice options modulate the subcortical neural activity that ultimately results in the decision, made below the level of conscious awareness, of moving your foot leftward to the brake pedal or rightward to the gas pedal. The hierarchy of decision making is adaptive: It would be unwieldy and time-consuming to make every decision consciously, and the brain has evolved to make many low-level decisions subconsciously.

In parallel with the hierarchy of decisional processes in the brain, decision making can be studied at a range of spatial and temporal scales (Sejnowski et al., 2014). At the most fundamental level, sub-second activity patterns of individual neurons can be measured and related directly to sensory or motor variables. In contrast, functional magnetic resonance imaging estimates the activity of a pooled groups of neurons over a period of several seconds in order to examine how activity is coordinated across several brain areas. Finally, sociologists study collective decision making that arises within and among groups of people, informing our understanding of behavior in crowds and market forces (Firdaus & Zusmelia, 2023). Each of these complementary scales enriches our understanding of how the brain mediates decision making.

Several questions examined in subsequent chapters relate to the model of hierarchical decision making. First, observe that the cortical regions responsible for your decision at the traffic light (e.g., "speed up") are just as constrained by the laws of nature as the subcortical regions responsible for directing your foot (e.g., "rightward"). The cortical neural activity representing "speed up" arose from activity in upstream neurons, which also follow the laws of nature. How, then, is conscious decision making compatible with the deterministic nature of brain function? This question is the focus of a long and rich philosophical debate, and Chapter 3 ("Neuroscience and the Freewill Debate") examines what neuroscience can contribute.

Second, in the legal context, the hierarchical organization of decision-making calls traditional conceptions of intent and culpability into question. While cortical activity might drive intentional decisions, subcortical processes can influence impulsive choices, leaving uncertainty about the degree of conscious control. This interplay between different levels of processing is a cornerstone to modern debates about *mens rea* ("guilty mind") and the viability of defenses like "My brain made me do it". Chapters 5 and 6 ("Neurolaw and Psychiatry" and "Stochastic Determinism and Criminal Law") explore how neuroscience informs our understanding of the conscious and unconscious contributions to decision making within the legal system.

Finally, decisions at all levels of the hierarchy are subject to external influences, even when they are experienced as being one's own. For example, advertisements and store layouts influence which products we purchase, a phenomenon explained by behavioral economics (Thaler, 2016). The field of behavioral public policy attempts to leverage such influence to "enhance" decision making by promoting decisions that benefit the individual and society (Thaler & Sunstein, 2008). How do these policies influence decision making, and what can neuroscience add beyond behavioral economics? These questions are addressed in Chapter 2 ("Behavioral Economics and Neuroeconomics: A Brief History and Overview").

1.2 Decision Making Requires Integrating Information across Levels

Decisions are not made in isolation but are instead influenced by myriad factors. As noted, even decisions that we think of as our own are actually influenced by external sources, for good or ill. Internal top-down influences, representing prior experiences and beliefs, also play an important role in decision making. To take a simple example, we can differentiate between a small nearby object and a large distant object, even if they are identically represented at early processing stages of the visual system, based on our prior knowledge of the size of the object (e.g., we know that it is a large airplane flying far away rather than a tiny airplane nearby). Prior experiences and beliefs are thought to be represented in patterns of activity in prefrontal cortex, which can then be integrated with new external information – such as current sensory information – to influence decisions. As we continuously interact with and make decisions about the world, we update these priors to optimize subsequent decisions (Ma, 2019). Priors relevant to higher-level decisions can be thought to represent our "worldview". Since worldviews can differ across individuals, people make different decisions even when presented with the same information.

This is particularly true when making ethical decisions, which may include competing priorities such as social norms, personal values, and immediate emotional responses. The mental processes required to balance these various considerations are subserved by integrating information across different levels and brain areas. How we navigate the maze of personal and societal considerations in making ethical decisions is discussed in Chapter 8 ("Introduction to Neuroethics").

Related, a feature of integrating information across hierarchical levels is that the expected outcomes of lower-level decisions can be evaluated at higher levels. The capacity for evaluation is particularly critical for ethical decision making. In Chapter 9 ("The Capacity for Evaluation of the Human Brain and Its Implications for an Artificial Moral Subject"), Farisco and Evers ask how this capacity bears on the moral status of artificially intelligent agents.

Religious experiences provide a rich arena in which to study how the brain integrates abstract, symbolic, sensory, and emotional information to derive meaning and purpose. Religion forms a key part of many individuals' worldview and serves as an overarching guide to ethical decision making. This worldview is further developed as individuals engage in practices aligned with their chosen faith. "The Neuroscientific Study of

Religious and Spiritual Phenomena" (Chapter 11) examines how brain activity can be shaped by activities such as prayer or meditation and how this activity might relate to one's subjective sense of well-being.

Humor demonstrates the brain's ability to integrate mismatched information into a coherent and emotionally satisfying resolution. The neural processes engaged by humor involve circuits for language processing, error detection, and reward. Higher brain areas process the discrepancy between expectation and surprise, while lower brain areas are involved in the emotional response. Integration across these levels gives rise to the gestalt we experience with the delivery of the punchline. Chapter 10 ("The Neuroscience of Humor") examines how the brain's ability to process humor facilitates social bonding and stress relief.

1.3 Conclusions

This introduction presents a brief review of the neuroscientific basis of decision making and provides a common conceptual framework for thinking about the relationship between neuroscience and the societally relevant topics discussed in subsequent chapters. We acknowledge some limitations of this treatment. First, our premise that neuroscience can offer any insight into societal topics, particularly when there are normative implications, is debatable. An argument from traditional moral philosophy is that empirical evidence can contribute nothing to normative questions: One cannot extrapolate from an "is" to an "ought" (Hume, 1739). Second, how the brain makes decisions is an active area of study and much remains to be discovered. New theories and experimental approaches are rapidly pushing the field forward, promising to shed yet more light on the topics discussed in this book and on other societally relevant questions.

Nevertheless, we hope to have convinced the reader to consider the subsequent chapters in light of the fundamental link between societal norms and brain function and to inspire the next generation of questions at the interface of neuroscience and society. In this journey, we may liken ourselves to Alice, wondering which way to go. The Cheshire Cat reminds us that this decision depends on our destination. Similarly, grappling with the interface between neuroscience and society begins by identifying the key goals and questions informing our exploration.

References

Bear, M. F., Connors, B. W., & Paradiso, M. A. (2016). *Neuroscience: Exploring the brain*. Wolters Kluwer.

Britten, K. H., Shadlen, M. N., Newsome, W. T., & Movshon, J. A. (1992). The analysis of visual motion: A comparison of neuronal and psychophysical performance. *The Journal of Neuroscience*, *12*(12), 4745–4765.

Donald, M. (1991). *Origins of the modern mind: Three stages in the evolution of culture and cognition* (pp. viii, 413). Harvard University Press.

Firdaus, F., & Zusmelia, Z. (2023). Recent research and theory on market in sociology: A semi-systematic literature review. *Jurnal Economia*, *19*(2), Article 2. https://doi.org/10.21831/economia.v19i2.48845

Gold, J. I., & Shadlen, M. N. (2007). The neural basis of decision making. *Annual Review of Neuroscience*, *30*, 535–574.

Greene, J. (2003). From neural "is" to moral "ought": What are the moral implications of neuroscientific moral psychology? *Nature Reviews Neuroscience*, *4*(10), 846–850. https://doi.org/10.1038/nrn1224

Herd, S., Krueger, K., Nair, A., Mollick, J., & O'Reilly, R. (2021). Neural mechanisms of human decision-making. *Cognitive, Affective, & Behavioral Neuroscience*, *21*(1), 35–57. https://doi.org/10.3758/s13415-020-00842-0

Luo, L., Callaway, E. M., & Svoboda, K. (2018). Genetic dissection of neural circuits: A decade of progress. *Neuron*, *98*(2), 256–281. https://doi.org/10.1016/j.neuron.2018.03.040

Ma, W. J. (2019). Bayesian decision models: A primer. *Neuron*, *104*(1), 164–175. https://doi.org/10.1016/j.neuron.2019.09.037

Miller, E. K., & Cohen, J. D. (2001). An integrative theory of prefrontal cortex function. *Annual Review of Neuroscience*, *24*, 167–202. https://doi.org/10.1146/annurev.neuro.24.1.167 24/1/167 [pii]

Newsome, W. T., Britten, K. H., & Movshon, J. A. (1989). Neuronal correlates of a perceptual decision. *Nature*, *341*(6237), 52–54. https://doi.org/10.1038/341052a0

Okazawa, G., & Kiani, R. (2023). Neural mechanisms that make perceptual decisions flexible. *Annual Review of Physiology*, *85*, 191–215. https://doi.org/10.1146/annurev-physiol-031722-024731

Salzman, C. D., Murasugi, C. M., Britten, K. H., & Newsome, W. T. (1992). Microstimulation in visual area MT: Effects on direction discrimination performance. *Journal of Neuroscience*, *12*(6), 2331–2355. https://doi.org/10.1523/JNEUROSCI.12-06-02331.1992

Sejnowski, T. J., Churchland, P. S., & Movshon, J. A. (2014). Putting big data to good use in neuroscience. *Nature Neuroscience*, *17*(11), 1440–1441. https://doi.org/10.1038/nn.3839

Shadlen, M. N., & Kiani, R. (2013). Decision making as a window on cognition. *Neuron*, *80*(3), 791–806. https://doi.org/10.1016/j.neuron.2013.10.047

Smith, P. L., & Ratcliff, R. (2004). Psychology and neurobiology of simple decisions. *Trends in Neurosciences*, *27*(3), 161–168. https://doi.org/10.1016/j.tins.2004.01.006

Steinmetz, N. A., Zatka-Haas, P., Carandini, M., & Harris, K. D. (2019). Distributed coding of choice, action and engagement across the mouse brain. *Nature*, *576*(7786), Article 7786. https://doi.org/10.1038/s41586-019-1787-x

Tekriwal, A., Afshar, N. M., Santiago-Moreno, J., Kuijper, F. M., Kern, D. S., Halpern, C. H., Felsen, G., & Thompson, J. A. (2019). Neural circuit and clinical insights from intraoperative recordings during deep brain stimulation surgery. *Brain Sciences*, *9*(7), Article 7. https://doi.org/10.3390/brainsci9070173

Thaler, R. H. (2016). Behavioral economics: Past, present, and future. *American Economic Review*, *106*(7), 1577–1600. https://doi.org/10.1257/aer.106.7.1577

Thaler, R. H., & Sunstein, C. R. (2008). *Nudge: Improving decisions about health, wealth, and happiness*. Yale Univ Pr.

2

Behavioral Economics and Neuroeconomics: A Brief History and Overview

Karolina M. Lempert

2.1 Introduction

When I enrolled in my first economics class in college, I flipped open my textbook, *Principles of Economics*, and read the following on the very first page: "Economists therefore study how people make decisions" (Mankiw, 2000). As a neuroscience major, I was taken aback by this statement. Surely, understanding how people make decisions was the domain of psychologists and neuroscientists. How could a discipline that I thought was all about money actually be about human beings?

It is clear to me now, though, that *several* disciplines are dedicated to elucidating human behavior, which is both complicated and fundamentally fascinating. Researchers in economics, psychology, neuroscience, political science, sociology, anthropology, cognitive science, and other fields have all attempted to understand human decision making from different perspectives or at different levels. Sometimes those perspectives are too dissimilar, making interdisciplinary work too difficult. In this chapter, however, I will discuss the developments in the twentieth century that have paved the way for the introduction of two interdisciplinary fields that both aim to advance theories of choice: behavioral economics (which combines psychology and economics) and neuroeconomics (which combines neuroscience, psychology, and economics).

One barrier to doing interdisciplinary research is that different fields use different terms. Therefore, researchers who want to do this kind of work should begin by understanding "natural kinds" within disciplines and then linking them across disciplines (Camerer, 2013). One such "natural kind" within economics is the concept of *utility*, which essentially refers to the satisfaction that someone might get from a state of the world. Psychologists and neuroscientists would not use the term "utility," but they have a term that means basically the same thing: *subjective value*. To the extent that someone desires something, it has subjective value to them. The fact that economists and neuroscientists agree that utility in economics is analogous to subjective value in neuroscience already

DOI: 10.1201/9781032717791-2

facilitates interdisciplinary connections between these fields (Glimcher and Rustichini, 2004). It took many years for these connections to solidify, though.

This chapter will be structured as follows. First, I will give an overview of how behavioral economics emerged as a discipline.[1] The field emerged largely because it became clear in the mid-twentieth century that economic models of choice did not capture actual human behavior. I will summarize the major findings and the key players that inspired behavioral economics. I will briefly mention a few major insights that have emerged from behavioral economics, and I will discuss how these insights have been applied so far. Then, I will summarize how neuroeconomics emerged and how it differs from behavioral economics. I will also describe a handful of the early seminal studies in neuroeconomics and discuss how insights from neuroscience might be applied in the real world as well.

2.2 The Quest to Reveal Utility (1700s–1930s)

It could be argued that the first time that psychology was brought into economics was with the introduction of the concept of utility. This concept can be illustrated with the following scenario:

> Imagine that a beggar finds a lottery ticket. There is a 50% chance that the ticket is worth $100,000. A rich man offers the beggar $30,000 for the ticket. Should the beggar sell the ticket to the rich man for $30,000?

The intuitive answer to this question is: of course! It seems unwise for a beggar with very little money to take a chance if he can have a nice chunk of money for sure. This intuitive answer is hard to justify mathematically, though. One simple algorithm that the beggar might use to make this decision is to take the *expected value* of keeping the lottery ticket and compare it to the $30,000 offer. To do this, the beggar would multiply the potential outcome of the lottery ($100,000) by the chances of winning it (0.5) and see if that value ($50,000) is higher or lower than $30,000. Since the expected value of keeping the lottery ticket is higher than the expected value of selling it, the beggar probably *should* keep the ticket, but, in reality, he almost certainly would not.

Economists were puzzled by scenarios such as this one. If people made their decisions by multiplying the *objective* value of an outcome (e.g., its monetary worth) by its probability, then (1) people would never make choices that led to a lower expected value (but they do), and (2) there would be no individual differences in choice (but there are). Daniel Bernoulli was able to explain these phenomena in 1738 when he proposed that people do not maximize money, or *objective* value; they maximize utility, or *subjective* value. According to Bernoulli (1738/1954):

[1] This chapter will borrow heavily from the introduction to the second edition of the *Neuroeconomics* textbook written by Glimcher and Fehr (2014), as well as from Camerer and Loewenstein (2004).

> The price of the item is dependent only on the thing itself and is equal for everyone; the utility however, is dependent on the particular circumstances of the person making the estimate . . . a gain of one thousand ducats is more significant to a pauper than to a rich man.
>
> **(p. 24)**

Bernoulli further argued that the relationship between a monetary gain and its utility is concave: the utility provided by each *marginal* gain *diminishes* as wealth increases. The shape of this function can explain why someone at a lower wealth level (like the beggar) is likely to be more risk averse than someone at a higher wealth level (like the rich man). Bernoulli's *expected utility theory* therefore built on expected value theory by saying that people maximize utility instead of value, and utility is just a mathematical transformation of objective value. By proposing a mathematical function for utility, Bernoulli had not only found a way to keep the mathematics in economics, but he had also (unwittingly) brought psychology into economics by recognizing that people might differ in what they find valuable.

Since Bernoulli introduced it, utility has become the "common currency" of economic decision making. All sorts of actions and outcomes can provide utility to a person, and the anticipated utility of those actions and outcomes is the main quantity that matters for choice. The dream, then, for many of the early economists, was to be able to measure utility directly. Francis Edgeworth (1845–1926) believed it would one day be possible to directly measure human desire through a "hedonimeter" (Colander, 2007). Edgeworth was inspired by developments in a new subfield of psychology at the time, psychophysics. Psychophysicists such as Ernst Weber (1795–1878) systematically mapped the relationships between external stimuli (such as weight, light, and sound) and physical sensations. They then came up with mathematical functions, such as the Weber–Fechner law, to describe those relationships. Edgeworth wanted to do something similar with human *desire*, measuring it in response to small changes in external stimuli (Colander, 2007). Edgeworth may have even noticed the striking resemblance between Bernoulli's (theoretical) utility/wealth curve and Weber's (empirical) stimulus/sensation curve.

Enthusiasm for Edgeworth's "hedonimeter" proposal waned quickly, however, for both practical and theoretical reasons. On the practical side, many economists of the time believed that measuring utility directly was simply not possible. Alfred Marshall wrote in 1890 that "It cannot be too much insisted that to measure directly, or *per se*, either desires or the satisfaction which results from their fulfilment is impossible, if not inconceivable" (1997, Ch. 3). After all, functional magnetic resonance imaging (fMRI) and other technologies that allow for the non-invasive measurement of human brain activity were still about a century away, and even now, these tools are expensive and noisy.

Many economists objected to Edgeworth's "hedonimeter" idea on different grounds, though: They thought that utility need only be measured indirectly, through the analysis of choice (Colander, 2007). Irving Fisher (1867–1947) argued that individuals reveal utility through their actions. Thus, by knowing the observable quantities (i.e., the stimuli that people are exposed to and the choices that they make about those stimuli), economists could *infer* the unobservable quantity of utility. In other words, there was simply no reason to measure utility directly as long as choices were observable. Fisher was so against the application of psychophysics for the study of utility that he wrote, "This foisting of Psychology on Economics seems to me *in*appropriate and vicious" (1892, p. 11).[2]

Most economists at the time agreed with Fisher. They began to think of utility as something that can be derived from choice and not measured separately from choice. Although Fisher, Arrow, Debreu, and others all had similar perspectives (Glimcher and Fehr, 2014), Paul Samuelson (1937) is often credited with formalizing the "revealed preference" approach in economics, and Von Neumann and Morgenstern (1944) are known for extending it into the domains of risky choice and game theory. What made the revealed preference approach so appealing is that, as long as individuals obeyed some consistency axioms, their behavior could be represented by the maximization of some utility function, implying that they behave *as if* they are maximizing some utility function (Glimcher and Fehr, 2014). Utility functions could be estimated from people's past choices and used to predict future choices. Measuring human desire did not seem necessary.

Of course, the assumption that people obey some consistency axioms is a critical one for the revealed preference approach and for expected utility theory. This assumption turned out to be a bad one, though. The discovery that people routinely violated these axioms led to the founding of behavioral economics.

2.3 Problems with Expected Utility Theory (1950s–1970s)

While going into the mathematical details of axioms of expected utility theory is beyond the scope of this chapter, I will describe a few axioms in simple terms. The most important axiom dictates that choices be *transitive*. For example, if someone prefers a Snickers bar over a Twix bar, and if they prefer Twix over Butterfinger, then they should prefer Snickers over Butterfinger. Transitivity is an important axiom for revealed preference theorists, because if choices are not transitive, then there is no mathematical utility function that can account for them. Simply put, there is no numerical utility value that one can assign to Snickers that is somehow both greater than the utility of Twix and less than the utility of Butterfinger at the same time. People are generally transitive in their choices; instances of intransitivity happen most often when preferences are not very strong.

[2] In fairness to Fisher, he did admire *some* psychologists: the behaviorists. Behaviorists, such as Watson and Skinner, concentrated on understanding the relationship between external stimuli and behavior, and they treated the mind like it was a "black box," not to be opened. The focus on the directly measurable was something that behaviorists had in common with revealed preference economists.

Under expected utility theory, choices should also be *description invariant*, meaning that people's choices should not depend on how choices are framed or presented.[3] Under the description invariance axiom, for example, if you prefer pizza to a hamburger for lunch, then it should not matter whether pizza is listed first or second on the menu.

Finally, choices should be *independent of irrelevant alternatives*. This "independence" axiom states that if an individual strongly prefers option A to option B, then that preference ordering should generalize, even to situations where there are more alternatives available (e.g., C, D, etc.). So, if you prefer a small coffee to a medium coffee, you shouldn't change your mind and pick the medium coffee just because a large coffee is added to the menu.

Humans violate the description invariance and independence axioms often. These axioms are built on the premise that preference orderings are stable over time, but it is more likely that preferences are constructed at the time of choice and that this makes them sensitive to context. The most famous early example of an independence axiom violation is now known as the Allais paradox (Allais, 1953). It was introduced by Maurice Allais at an economics conference in France in 1952. In one version of Allais's paradox, people are presented with two choice problems:

Problem 1:

Option 1A: $1 million [100% chance]

Option 1B: $1 million [89% chance] + $5 million [10% chance] + $0 [1% chance]

Most people prefer option 1A in this scenario.

Problem 2:

Option 2A: $1 million [11% chance] + $0 [89% chance]

Option 2B: $5 million [10% chance] + $0 [90% chance].

Here, most people prefer option 2B.

Although it is probably not immediately obvious, according to expected utility theory, if you prefer option 1A, then you *must* also prefer option 2A. That is because, once you "cancel out" the outcome that Options 1A and 1B have in common in Problem 1, the choice reduces to Problem 2. Note that Problem 1 can also be presented this way:

Option 1A: $1 million [11% chance] + *$1 million [89% chance].*

Option 1B: $5 million [10% chance] + $0 [1% chance] + *$1 million [89% chance]*

Here the common consequence is italicized: there is an 89% chance of winning $1 million either way. That common consequence should be irrelevant to people's choices, so the choice should depend only on whether you prefer $1 million with an 11% chance or $5

[3] They should also be "procedure invariant," meaning that the method by which a preference is *elicited* should not affect the preference, either.

million with a 10% chance. In Problem 2, the common consequence is $0 with an 89% chance:

Option 2A: $1 million [11% chance] + *$0 [89% chance].*

Option 2B: $5 million [10% chance] + $0 with a 1% chance + *$0 [89% chance].*

When the common consequence is $0 instead of $1 million, people change their minds about what they prefer. Why is that? In Problem 1, the common consequence makes it so that there is an option with a *guaranteed* monetary gain: Option 1A. In Problem 2, neither option offers a guarantee. *Certainty* appears to be special here.

Soon after Allais proposed his paradox, Daniel Ellsberg (1961) put forward another in which people violated the description invariance axiom. The reactions to the Allais and Ellsberg paradoxes were mixed. Some economists began working out the boundary conditions under which expected utility theory held. They believed that expected utility theory was still robust, and the paradoxes were just special cases, or exceptions. For example, Allais's paradox could be accounted for if expected utility theory held only in cases where there were no guaranteed options. Other scientists, like the political scientist Herbert Simon, acknowledged that consistency axiom violations occurred often, but they did not think that these violations meant that humans were irrational. Simon suggested that human rationality was "bounded" by the computational constraints of the human brain.

By far the most influential "reaction" to the paradoxes, though, was the research done by Daniel Kahneman and Amos Tversky in the late 1970s and 1980s.

2.4 Early Insights from Behavioral Economics (1970s–2000s)

2.4.1 Reference Dependence

Daniel Kahneman (1934–2024) and Amos Tversky (1937–1996) were both psychologists. They began working together to understand judgment and decision making in the late 1960s (Lewis, 2017). One of the reasons that Kahneman and Tversky were so influential was that they emphasized the distinction between *normative theories* and *descriptive theories* of human behavior. A normative theory is a theory of how people *should* behave, whereas a descriptive theory is a theory of how people *actually* behave. Expected utility theory is a normative theory of behavior. Having preferences that are transitive and consistent over time is normative. Expected utility theory was often applied as if it were a descriptive theory, though. As Allais's and Ellberg's paradoxes made clear, however, people simply do not follow the tenets of expected utility theory, so it is not a good descriptive theory.

A good economic theory is said to have generality, tractability, and congruence with reality (Camerer and Loewenstein, 2004). Expected utility theory is general and tractable, but it is not congruent with reality. The search was on, then, for a new theory (or set of theories) of choice that did actually explain the many ways in which people violated assumptions of expected utility theory. If this theory were general and tractable, that would be even better.

Kahneman and Tversky not only uncovered many biases and anomalies in human decision making through clever experiments, but they also proposed a formal, descriptive theory of decision making under uncertainty known as *prospect theory* and published their theory in one of the leading economics journals, *Econometrica* (Kahneman and Tversky, 1979). The fact that a theory put forward by psychologists was taken seriously by economists is exactly why Kahneman and Tversky are often credited with founding the field of behavioral economics, a subfield of economics. It is also why Kahneman was awarded the Nobel Prize in Economics in 2002.[4]

Prospect theory, like expected utility theory, is a theory of how people make risky choices, or choices where at least one of the outcomes is uncertain. In expected utility theory, these choices are thought to be made by multiplying the probability of a prospect by its utility to get its expected utility. The option with the highest expected utility is the one that *should* be chosen (remember: expected utility theory is a *normative* theory). With prospect theory, Kahneman and Tversky adapted expected utility theory to better account for empirical data. Most importantly, they introduced the psychophysical idea of *reference dependence* into the model. They wrote the following:

> When we respond to attributes such as brightness, loudness, or temperature, the past and present context of experience defines an adaptation level, or reference point, and stimuli are perceived in relation to this reference point. . . . The same principle applies to non-sensory attributes such as health, prestige, and wealth.
>
> **(Kahneman and Tversky, 1979, p. 277)**

In other words, an important feature of decision making is that subjective value depends on whether an outcome is perceived as a loss or a gain relative to some reference point. People have different risk attitudes when considering potential gains than they do when facing potential losses. This phenomenon was demonstrated most memorably in Kahneman and Tversky's "Asian disease" problem (Tversky and Kahneman, 1981):

Problem 1:

Imagine that the U.S. is preparing for the outbreak of an unusual Asian disease, which is expected to kill 600 people. Two alternative programs to combat the disease have been proposed. Assume that the exact scientific estimate of the consequences of the programs are as follows:

[4] Tversky would have shared the prize had he been alive at the time. Amos Tversky died in 1996, and Nobel Prizes are not awarded posthumously.

Option 1A: If Program A is adopted, 200 people will be saved.

Option 1B: If Program B is adopted, there is a 1/3 probability that 600 people will be saved, and a 2/3 probability that no people will be saved.

Consistent with the fact that people tend to be risk averse for potential gains, 72% of respondents preferred Program A (Option 1A). However, what happens when people are given the same prompt but with the following options instead?

Problem 2:

Option 2A: If Program C is adopted, 400 people will die.

Option 2B: If Program D is adopted, there is a 1/3 probability that nobody will die, and a 2/3 probability that 600 people will die.

With this framing, the vast majority of participants (78%) were risk *seeking*, selecting Option 2B.

Integrating a reference point into the prospect theory model allowed Kahneman and Tversky to explain the fact that people tended to be risk seeking when considering potential losses and risk averse when considering potential gains. In prospect theory, the utility function for perceived gains is concave (just as Bernoulli suggested), but the utility function for perceived losses is actually *convex*.[5] The subjective value of potential gains and losses changes most drastically when the changes are close to the reference point.

Importantly, reference dependence can be demonstrated even when losses and gains are not real but rather just perceived. In fact, the two problems in the "Asian disease" paradox have the exact same outcomes; they are just framed differently. Since 600 people are expected to die in the disease outbreak, Program A is equivalent to Program C. It is *implied* that if 200 people are saved, 400 people will die. Moreover, Program B is the same as Program D: a 1/3 chance of 600 people being saved is the same as a 1/3 chance of no one dying. In the first version of the "Asian disease" problem, the potential change is perceived as positive (i.e., a gain): the opportunity to save lives. Diminishing marginal utility in the domain of *gains* predicts risk aversion in the prospect theory framework. In Problem 2, the potential change is perceived as negative (i.e., a loss): lost lives. Diminishing marginal utility in the domain of *losses* predicts risk seeking in the prospect theory framework.[6] Expected utility theory would not be able to explain people's choices in these two problems, because their behavior is in violation of the description invariance axiom.

[5] Kahneman and Tversky were not the first to suggest that utility functions differ for potential losses compared to gains. Although he has not received as much attention, Markowitz(1952) also discussed these ideas.

[6] Prospect theory also proposed a function that related objective probabilities to subjective "decision weights" to account for the fact that people overweight very small probabilities. For the sake of space, I will point readers to Kahneman and Tversky's seminal 1979 paper for more details about this piece of the model.

2.4.2 Loss Aversion

There is another feature of the prospect theory value function that is critical for explaining people's decisions. Not only do people have differing risk attitudes for gains compared to losses, but the anticipated change in utility that comes from a potential loss is more extreme than the change in utility brought on by a potential gain. Simply put, people dislike losing more than they like winning. The way that loss aversion is incorporated into the prospect theory value function is by making the utility function for losses steeper than the utility function for gains.

There is a great deal of evidence that people are loss averse. Loss aversion explains why most people will not take a fair bet on a coin flip – if you offer someone a 50/50 chance to win $5 or lose $5, most people will refuse that offer. Loss aversion can also explain the endowment effect, which is the phenomenon that sell prices are higher than buy prices. Daniel Kahneman, Jack Knetsch, and Richard Thaler (1990) did a series of experiments to demonstrate the endowment effect in a lab setting. They randomly endowed half of the participants with coffee mugs. Those who had mugs were asked for the lowest price at which they would sell their coffee mug. Those who did not have mugs were asked for the lowest price at which they would buy a coffee mug from one of the "seller group" participants. Although all participants were aware that the mug assignments were random, and the group that "owned" the mugs had only owned them for a few minutes, selling prices (median = $5.75) were about twice as high as buying prices were (median = $2.25). Loss aversion can explain this result. Since the pain of losing a mug is greater than the pleasure of gaining a mug, participants who had a mug to lose demanded greater compensation in order to sell the mug.

One limitation of this study was that it was not clear whether the result was driven by the sellers experiencing loss aversion for the *mug* or whether it was driven by the buyers experiencing loss aversion for *money*. It is possible that the buyers might have been offering lower buy prices because the pain of losing money (for the buyers) is greater than the pleasure of getting money (for the sellers). In a clever follow-up study, Kahneman and colleagues (1990) introduced a third group: "choosers," who simply decided whether, for a series of prices, they would prefer to have a mug or to have money. The "choosers" in the follow-up experiment are a better control group than the "buyers" in the first experiment, because they are not in a position of potentially losing their own money to buy a mug. The median choosing price turned out to be half of the median selling price in that experiment ($3.12 and $7.12, respectively). Therefore, it is clear that just owning an item, even temporarily, imbues that item with extra value, since losing the item will elicit a greater "utility hit" than the gain of another item would. The endowment effect is one of several *anomalies*, or deviations from economically rational behavior, discovered by Richard Thaler, who, along with Kahneman and Tversky, is considered one of the founders of behavioral economics[7] (Thaler, 1980; Kahneman et al., 1991).

[7] For his work linking psychological principles (such as loss aversion) to economic phenomena, he won the Nobel Prize in Economics in 2017.

The endowment effect is one corollary of loss aversion. Another is the status quo bias (Samuelson and Zeckhauser, 1988). According to the status quo bias, people tend to disproportionately stick with the "default" option when they make decisions. The default option can be demarcated by someone else (like an experimenter), or it can simply be whatever choice had already been made in the past. The status quo bias is there even if there is no good reason one of the options had been pre-selected. For example, people tend to stick with the electricity service they already have, even if it is unreliable (Hartman et al., 1991). People tend to stick with health plans and retirement plans that have been arbitrarily chosen for them (Samuelson and Zeckhauser, 1988). Status quo bias is thought to be related to loss aversion by the following logic. People feel that the default option is somehow *theirs*, because they picked it before or because it was picked for them. Therefore, an alternative option would have to be much better than the default option in order for people to agree to "lose" the option that they already have.

Some critics point out that the default option is often something that the person has no experience with, so it would be a stretch to say that people feel as if they "own" it. But recent research in behavioral economics has shown that reference points are set not only by what you currently own; they may actually be set by your *expectations* of what you will own (Köszegi and Rabin, 2006, 2007). For example, a year-end bonus of $10,000 is technically a gain, but if you expected a $20,000 bonus, it will feel like a loss. If reference points are set by expectations in this way, then the applications of loss aversion are even broader than what was previously thought. How reference points emerge and how different kinds of reference points are integrated are active areas of study in behavioral economics. For example, another descriptive theory of decision making under uncertainty that has recently been proposed represents the reference point as a weighted average of past outcomes (Glimcher and Tymula, 2023).

2.5 Behavioral Economics in the Field (2000s–present)

Loss aversion and its corollaries have been extremely influential in the field of behavioral economics. First of all, economic phenomena that were previously puzzling could be explained by referencing loss aversion (Camerer and Loewenstein, 2004). For example, one study done by Genesove and Mayer (2001) that focused on the real estate market found that sellers listed their homes at prices that were strongly affected by the prices at which they originally bought their homes. Since the purchase price of the home is a sunk cost, the best strategy for choosing a list price is to anticipate what the market will pay. However, it seems as if sellers do not want to "lose" money on a sale, even though it would be a loss in name only. Another example is the asymmetry in consumers' responses to product price changes (Hardie et al., 1993). Consumers' increase in purchasing following a price drop is not as extreme as their decrease in purchasing following a price increase. Essentially, people are responsive to changes, and changes that result in a loss are more salient than those that result in a gain.

Another way in which loss aversion has been influential is that it is often at the heart of "nudges" (Thaler and Sunstein, 2003, 2008). Richard Thaler and Cass Sunstein defined nudges as small changes to a choice's framing, or "choice architecture," that alter people's decisions in systematic ways without forbidding options or changing economic incentives (Thaler and Sunstein, 2008). Nudges can be used by institutions that offer choices to people (e.g., governments, banks) to try to change people's decisions in a manner that is subtle but often highly effective.

A perfect example of a nudge is to label one of the choice options as the "default" option. This makes it more likely that people will choose that option – due to status quo bias – without making the other options unavailable. Nudges that are based on changing defaults are powerful. For example, in countries where people have to "opt-in" or give explicit consent to be organ donors, organ donation rates are low (~24%). In countries where explicit consent is needed only to opt *out* of being an organ donor, organ donation rates are much higher – nearly 100% (Johnson and Goldstein, 2003). Defaults have also been used to increase retirement savings, since people can be defaulted into making contributions to their retirement plans at work (Thaler and Benartzi, 2004; Carroll et al., 2009).

Defaults are not the only kinds of nudges. For example, vaccine uptake rates are higher when people are assigned appointment times (Chapman et al., 2010) or are nudged to make a concrete plan to get a vaccine (Milkman et al., 2011). When people are given information about their energy use and how it compares to their neighbors', they cut back on their electricity use (Allcott, 2011). Even just streamlining the process of applying for financial aid makes people more likely to go to college (Bettinger et al., 2012). One of the reasons nudges are so exciting in the field of public policy is that they are often low cost. Changing how questions are worded is a much cheaper strategy for altering decision making compared to providing tax incentives or changing prices, and nudges can be as effective or more effective than those more expensive strategies (Benartzi et al., 2017).

Nudges are controversial, too, though. Some scholars believe that it is unethical to knowingly attempt to change people's choices. People generally accept that they will be manipulated by companies that are trying to make a profit, but they may find it harder to stomach the thought of governments manipulating them (e.g., by making organ donation a default option). However, defenders of nudging rightly point out that all choices have some sort of architecture. Anyone who poses a choice to someone else must make a decision about how to frame that choice. So, why *not* frame the choice in such a way that is generally beneficial?[8] Proponents of nudging also emphasize that nudges do not actually prevent people from choosing what they want.

Another reason nudges are controversial is that sometimes they are ineffective, or they can even backfire (e.g., Holzmeister et al., 2022) or have unintended effects on other outcomes. Therefore, it is important to do extensive small-scale testing on any new

[8] Of course, this becomes tricky when incentives between institutions and individuals are misaligned.

nudge strategy before implementing it widely. In addition, different kinds of nudges have different success rates. Those that require less effort on the part of the chooser have larger effect sizes (Beshears and Kosowsky, 2020).

2.6 Behavioral Economics Summary

Behavioral economics came into its own as a subfield of economics in the 1970s and 1980s. It aims to understand the limits of human rationality in decision making, to codify those limits, and to explore their empirical implications (Glimcher and Fehr, 2014). One way to summarize the approach of behavioral economists is the following. First, they identify anomalies in choice by comparing human behavior to the behavior predicted by normative theories, such as expected utility theory. Then, they create alternative theories that generalize existing theories. Finally, they derive the implications of the new, alternative models and test them (Camerer and Loewenstein, 2004). In essence, behavioral economists incorporate psychological principles into economic theories of choice. The work on reference dependence and loss aversion that was done by Kahneman, Tversky, Thaler, and others has had the broadest impact so far. However, new descriptive theories of human choice have been emerging since then in multiple domains. Psychological principles have been applied to models of social preference (Camerer, 2003) and even to models of the stock market. "Behavioral finance" is a subfield of behavioral economics that deals specifically with understanding the decision making of investors and financial professionals (Shleifer, 2000). Overall, the collaborations between psychologists and economists that have emerged over the last fifty years have been fruitful, and behavioral economics is poised to continue expanding.

2.7 From Behavioral Economics to Neuroeconomics (1990s–2000s)

The goal for behavioral economists is roughly the same as that of traditional, neo-classical economists; they want to be able to predict people's choices. Recall from earlier in the chapter that, in the early 1900s, economists had abandoned the idea of measuring utility directly, and they instead decided to take a "revealed preference" approach, wherein utility was inferred from choice. With this approach, past choices should be sufficient to predict future choices, even in the absence of any mechanistic knowledge about how exactly decision making is implemented. To this day, many economists do not care about the neural mechanisms underlying choice, because they do not believe that this knowledge will offer any advance in choice prediction (Gul and Pesendorfer, 2008). They contend that neuroscience and economics should remain independent, since they have fundamentally different goals.

However, there are situations in which past choices are not available, and prediction using economic models alone becomes difficult. For example, it is hard to know how much people will value brand new products, and it is hard to know how they will respond to unprecedented policy changes (Camerer, 2013). In these cases, understanding the mechanisms behind decision making – including the neural mechanisms – might be useful to economists. This is where *neuroeconomics* comes in.

Neuroeconomics is an interdisciplinary field that spans economics, psychology, and neuroscience (Glimcher and Rustichini, 2004). The goal of neuroeconomics is to "understand the processes that connect sensation and action by revealing the neurobiological mechanisms by which decisions are made" (Glimcher and Rustichini, 2004, p. 447), with the hope that a general theory of decision making will emerge that is mechanistic, behavioral, and mathematical (Glimcher and Rustichini, 2004). Psychological and neuroscientific theories tend to be mechanistic and behavioral but not mathematical, whereas economic theories tend to be mathematical and behavioral but not mechanistic (Camerer, 2013).

Neuroeconomics is a relatively new field, with the first neuroeconomic papers being published toward the very end of the 1990s and early 2000s. One reason neuroeconomics might have emerged around this time is because of the invention of functional magnetic resonance imaging circa 1992 (Bandettini et al., 1992; Ogawa et al., 1992). Functional MRI is a technology that allows researchers to crudely measure a person's neural state while they are engaging with some stimulus or task. Before the introduction of fMRI, the only way to non-invasively measure a proxy of neural activity in real time was with positron emission tomography (PET) imaging. PET imaging, however, requires radioactive tracers, has poor temporal resolution, and is very expensive, so its use did not become widespread in cognitive neuroscience. MRI scanners are widely available and well tolerated, though, so the early twenty-first century saw an explosion of fMRI studies. The collection of data outpaced the development of theory, though. Given that economics already had elegant, testable theories of choice available, both cognitive neuroscientists and behavioral economists alike started to measure fMRI signals during economic choice in order to test these economic theories.

2.8 Early Insights from Neuroeconomics (1990s–present)

2.8.1 Utility Signals in the Brain

There were several questions from both neoclassical and behavioral economics that neuroeconomists hoped they could answer. The first was: is there any evidence that utility is represented in the brain? Could an MRI machine serve as a "hedonimeter" of sorts? In 1999, Paul Glimcher, a neuroscientist at New York University, and his postdoc Michael Platt decided to do the kind of psychophysics-inspired experiment that Edgeworth dreamed of. They recorded neural activity from electrodes placed in the lateral intraparietal area of monkeys while those monkeys made decisions about juice rewards that varied in both magnitude (number of drops of juice) and probability (chances of getting juice). Monkeys made these decisions using eye movements. In expected utility theory, decisions under uncertainty are thought to be made by weighing the magnitude of rewards by the probability of getting those rewards. Therefore, if there are neural signals that increase as probability and magnitude increase in a multiplicative fashion, and if these neural signals predict choice, then they can be interpreted as utility signals. Platt and Glimcher (1999) did indeed find signals like these in the lateral intraparietal area of the monkey. This study is thus credited as being one of the first to find evidence of utility in the brain.

A few years later, Kable and Glimcher (2007) found evidence of utility in the human brain using fMRI. While lying in the MRI scanner, participants did an economic intertemporal choice task in which they made a series of choices between receiving $20 now and receiving a larger monetary reward after a delay (e.g., $30 in 1 week). The amounts and delays varied from trial to trial. People tend to discount the value of delayed rewards as delays increase, an idea that was first proposed by Samuelson (1937) in his discounted utility model. Kable and Glimcher (2007) used participants' choices combined with a model of intertemporal choice[9] to compute the subjective value, or utility, of each delayed reward for each participant in the study. Their goal was to see if activity in any part of the brain increased as subjective value increased. They found that both the ventral striatum and ventromedial prefrontal cortex carried these subjective value signals (Kable and Glimcher, 2007). Since then, this result has been replicated hundreds of times in different choice domains (for meta-analyses, see Bartra et al., 2013; Clithero and Rangel, 2014). The idea that the ventromedial prefrontal cortex in particular is causally involved in assessing subjective value or utility is also supported by lesion studies showing that people with damage to this area make intransitive choices (Fellows and Farah, 2007).

Kable and Glimcher's research was important not just because it demonstrated that subjective value was represented in the ventromedial prefrontal cortex but also because it showed that the ventromedial prefrontal cortex could represent the value of *both* immediate and delayed rewards (Kable and Glimcher, 2010). That particular finding challenged a behavioral economic model of intertemporal choice that suggested that immediate and delayed rewards were evaluated by separate neural systems (Laibson, 1997; McClure et al., 2004). It is beyond the scope of this chapter to go into the details of this debate. It is worthy of mention, however, since the idea that neural evidence could be used to arbitrate between economic models was groundbreaking at the time.

Some neuroeconomists, such as Glimcher, took neoclassical economics in general (and expected utility theory specifically) as a starting point in their neuroscientific studies, mostly because of the elegant axioms associated with it. Other neuroeconomists started by testing behavioral economic theories, such as prospect theory. For example, Breiter and colleagues (2001) conducted an fMRI study in which they varied the appeal of a $0

---

[9] Samuelson's discounted utility model proposes an exponential relationship between delay and utility. As the delay until a reward arrives increases, the utility of that reward decreases exponentially. Empirically, the relationship between delay and utility appears to be hyperbolic, though, rather than exponential. Thus, Kable and Glimcher fit a hyperbolic model in their study, even though that is not common practice in economics (for tractability reasons).

outcome in different blocks by making the counterfactual outcome either a reward (e.g., +$1) or a punishment (e.g., –$1). The idea behind this study was to find neural evidence for reference dependence. If the utility of $0 depended on whether it was perceived as a gain or a loss, then the $0 outcome should elicit greater reward signals in the ventral striatum, a part of the brain that had been previously shown to represent reward (Delgado et al., 2000), when the counterfactual outcome was a punishment compared to when the counterfactual outcome was a reward. That is what Breiter and colleagues found, thus providing physiological evidence that utility is reference dependent. Since then, many studies have shown evidence for reference-dependent subjective value signals.

2.8.2 Loss Aversion in the Brain

These early neuroeconomic studies focused mostly on the *measurement* of utility signals in different contexts. This was an important pursuit at the beginning, since it established that economic theories, principles, and methods could be combined with neuroscience. The main benefit of using neuroscience methods, though, is that they can help elucidate the *mechanisms* underlying choice. The discovery of loss aversion was revolutionary because it showed how psychology could be used to improve the predictive validity of economic models. The term loss aversion is merely descriptive, however. Knowing *that* people overweight negative outcomes compared to positive outcomes is not the same as knowing *why* they do so or knowing *how* that asymmetric weighting is implemented. The hope of neuroeconomics is that answering questions about this mechanism could improve models of choice even further.

Some work has been done to investigate the neural mechanisms of loss aversion, and emotion appears to play a key role. Emotion, which has typically been the domain of psychologists, is a notoriously difficult topic to study. Part of the reason for this is because scientists do not all agree on a definition of emotion (Barrett, 2006). Without a clear way to operationalize emotion, quantifying it is a daunting task. Given the heavy emphasis on quantitative predictions in economics, economists have thus typically ignored emotion in their work, or they have relegated it to just being a source of random "noise" in decision making. But what if emotion has a systematic role in decision making, rather than a random one?

Advances in both behavioral economics and affective neuroscience have helped scientists to start clarifying the role of emotion in decision making (for reviews, see Lempert and Phelps, 2013; Lerner et al., 2014; Phelps et al., 2014). For example, emotional *arousal* in response to a stimulus can actually be quantified. Sympathetic nervous system responses, or "fight-or-flight responses," such as sweating, pupil dilation, and increased heart rate, can be measured using non-invasive tools, including skin conductance and eye tracking. Therefore, it is possible to measure the emotional salience of economic choices and outcomes.

Studies of emotional arousal have uncovered evidence of reference dependence and loss aversion. Arousal depends not just on the value of a stimulus but rather on its value relative to an expectation (Lempert et al., 2015). Monetary losses evoke a greater emotional arousal response than gains do (Sokol-Hessner et al., 2009). Moreover, the brain region that is most involved in orchestrating emotional arousal responses, the amygdala, is more active when people experience monetary losses compared to gains (Tom et al., 2007; Sokol-Hessner et al., 2013). These studies are correlational, but there is some evidence that emotional arousal plays a causal role in loss aversion. People with amygdala lesions are less loss averse than healthy people (De Martino et al., 2006), and reducing arousal pharmacologically using beta-blockers reduced loss aversion (Sokol-Hessner et al., 2015).

Understanding the neural mechanisms of loss aversion and other behavioral economic phenomena may prove useful to economists for a few reasons. First of all, the degree to which people experienced emotional arousal following loss outcomes was associated with their loss aversion parameter, derived from a prospect theory–inspired model (Sokol-Hessner et al., 2009). This suggests that it may one day be possible to predict people's risk attitudes from knowledge of their emotional state. Second, emotion regulation techniques can be employed to reduce emotional responses, and that reduction can, in turn, alter loss aversion (Sokol-Hessner et al., 2009). Thus, knowledge about the emotional underpinnings of risky choice may also lead to the development of new nudges and other interventions to change choice.

2.9 Neuroeconomics in the Field?

Given that neuroeconomics is a young field, its insights have not yet been applied as broadly as those of behavioral economics have. One of the issues with using neuroscience in the field is that the methods of neuroscience, like fMRI, are expensive and require specialized knowledge to use. The applied field where neuroscience has caught on most is marketing. Business schools around the country, such as Wharton at the University of Pennsylvania, are welcoming neuroscientists to their faculty and offering neuroscience courses. Their hope is that as physiological data becomes easier to collect, these data may complement data collected through more traditional means (e.g., questionnaires, focus groups) to help with promoting products.

To convince people in business that it is worthwhile to adopt neuroscience methods, though, there must be some evidence that neural data can predict consumer preferences better than self-report can. Is there any? In one study by Berns and Moore (2012), a small focus group of ~30 adolescents was scanned while listening to new music clips. The adolescents also rated how much they liked each song. The authors then totaled the sales for each of these songs over the next three years. The subjective liking ratings of the focus group were not predictive of the songs' popularity. However, the songs that elicited a greater response in the ventral striatum – a reward region of the brain – were the ones that turned out to be the most commercially successful (Berns and Moore, 2012). This study was exciting because it suggested that even scanning just a small group of people may be sufficient to make predictions about the success of a product on a much larger scale. Scanning every consumer is impractical, but for large corporations, scanning a few people before launching a product may actually be beneficial, especially in domains where it is difficult to predict what will be popular.

Several other studies have now found that brain activity explains additional variance in choice, over and above what could be predicted by self-report measures alone (Knutson and Genevsky, 2018). Although some studies have found that activity in the ventromedial prefrontal cortex is predictive of future choice (e.g., responses to ads predicting ad-related calls; Falk et al., 2012), a consensus is emerging that the ventral striatum is the region in the brain where reward signals are *most* predictive of future choice (Genevsky et al., 2017).

These "neuro-forecasting" studies are promising, but neuroimaging remains a difficult tool to use in the field. There has been a general push toward incorporating other kinds of process data in order to improve marketing outcomes, though. For example, powerful eye trackers are now wearable, so marketing researchers can study people's eye movements while they are shopping in stores to see what information they are attending to. As the tools of neuroscience get cheaper and easier to use, they will be adopted more in the field. For now, however, neuroeconomics has not influenced policy or marketing nearly as much as behavioral economics has.

2.10 Conclusion

In some ways, psychology and neuroscience began to be integrated into economics in 1738 when Bernoulli proposed the idea of utility. Utility is a latent, unobservable variable that connects stimuli (e.g., a choice between two gambles) to action (e.g., choosing the safe option). In that sense, utility is both psychological (since it is created by the mind) and neurobiological (since the mind is instantiated in the brain). Measuring utility directly and objectively was deemed both impossible and unnecessary by the neoclassical economists of the early twentieth century. They devised clever axioms, and if people followed them, then their future choices could be predicted from their past choices.

The main impetus for behavioral economics was data: people routinely violated the axioms of expected utility theory. Therefore, expected utility theory was labeled a normative theory rather than a descriptive one. Behavioral economists now use the methods of psychology (e.g., experiments, surveys, tasks) to create better descriptive models of choice. The main impetus for neuroeconomics, on the other hand, was methodological: the introduction of new methods for measuring brain activity, such as fMRI, made it possible to measure utility directly. Neuroeconomists now use the methods of both neuroscience and psychology to propose descriptive models that are neurobiologically plausible.

Decision making has been studied by scholars in many different disciplines, including neuroscience, psychology, and economics. These fields have typically taken different approaches to studying human behavior, so until recently, they were completely independent from each other. Taking an interdisciplinary approach can be hugely beneficial, however. First of all, connecting "natural kinds" between disciplines (e.g., connecting utility to subjective value) can lead to novel and creative insights about those constructs. Second, different disciplines have mastered different tools, so it is efficient to try to combine those tools. Economists' specialty is constructing formal models of choice that make strong, falsifiable predictions. Psychologists and neuroscientists specialize in measuring the processes by which choices are made. By combining theories from economics with data from psychology and neuroscience, we may have a general, mathematical, mechanistic, and realistic theory of decision making one day soon.

References

Allais M (1953) Le Comportement de l'Homme Rationnel devant le Risque: Critique des Postulats et Axiomes de l'Ecole Americaine. *Econometrica* 21:503.

Allcott H (2011) Social Norms and Energy Conservation. *Journal of Public Economics* 95:1082–1095.

Bandettini PA, Wong EC, Hinks RS, Tikofsky RS, Hyde JS (1992) Time Course EPI of Human Brain Function During Task Activation. *Magnetic Resonance in Medicine* 25:390–397.

Barrett LF (2006) Are Emotions Natural Kinds? *Perspectives on Psychological Science* 1:28–58.

Bartra O, McGuire JT, Kable JW (2013) The Valuation System: A Coordinate-Based Meta-Analysis of BOLD fMRI Experiments Examining Neural Correlates of Subjective Value. *NeuroImage* 76:412–427.

Benartzi S, Beshears J, Milkman KL, Sunstein CR, Thaler RH, Shankar M, Tucker-Ray W, Congdon WJ, Galing S (2017) Should Governments Invest More in Nudging? *Psychological Science* 28:1041–1055.

Bernoulli D (1954) Exposition of a New Theory on the Measurement of Risk. *Econometrica* 22:23.

Berns GS, Moore SE (2012) A Neural Predictor of Cultural Popularity. *Journal of Consumer Psychology* 22:154–160.

Beshears J, Kosowsky H (2020) Nudging: Progress to Date and Future Directions. *Organizational Behavior and Human Decision Processes* 161:3–19.

Bettinger EP, Long BT, Oreopoulos P, Sanbonmatsu L (2012) The Role of Application Assistance and Information in College Decisions: Results from the H&R Block Fafsa Experiment*. *The Quarterly Journal of Economics* 127:1205–1242.

Breiter HC, Aharon I, Kahneman D, Dale A, Shizgal P (2001) Functional Imaging of Neural Responses to Expectancy and Experience of Monetary Gains and Losses. *Neuron* 30:619–639.

Camerer C (2003) *Behavioral Game Theory: Experiments in Strategic Interaction.* New York, NY: Russell Sage Foundation [u.a.].

Camerer CF (2013) A Review Essay About *Foundations of Neuroeconomic Analysis* by Paul Glimcher. *Journal of Economic Literature* 51:1155–1182.

Camerer CF, Loewenstein G (2004) Behavioral Economics: Past, Present, Future. In: *Advances in Behavioral Economics* (Camerer CF, Loewenstein G, Rabin M, eds), pp 3–51. Princeton, NJ: Princeton University Press.

Carroll GD, Choi JJ, Laibson D, Madrian BC, Metrick A (2009) Optimal Defaults and Active Decisions*. *Quarterly Journal of Economics* 124:1639–1674.

Chapman GB, Li M, Colby H, Yoon H (2010) Opting In vs Opting Out of Influenza Vaccination. *JAMA* 304:43.

Clithero JA, Rangel A (2014) Informatic Parcellation of the Network Involved in the Computation of Subjective Value. *Social Cognitive and Affective Neuroscience* 9:1289–1302.

Colander D (2007) Retrospectives: Edgeworth's Hedonimeter and the Quest to Measure Utility. *Journal of Economic Perspectives* 21:215–226.

Delgado MR, Nystrom LE, Fissell C, Noll DC, Fiez JA (2000) Tracking the Hemodynamic Responses to Reward and Punishment in the Striatum. *Journal of Neurophysiology* 84:3072–3077.

De Martino B, Kumaran D, Seymour B, Dolan RJ (2006) Frames, Biases, and Rational Decision-Making in the Human Brain. *Science* (New York, NY) 313:684–687.

Ellsberg D (1961) Risk, Ambiguity, and the Savage Axioms. *The Quarterly Journal of Economics* 75:643.

Falk EB, Berkman ET, Lieberman MD (2012) From Neural Responses to Population Behavior: Neural Focus Group Predicts Population-Level Media Effects. *Psychological Science* 23:439–445.

Fellows LK, Farah MJ (2007) The Role of Ventromedial Prefrontal Cortex in Decision Making: Judgment Under Uncertainty or Judgment per se? *Cerebral Cortex* 17:2669–2674.

Fisher I (1892) *Mathematical Investigations in the Theory of Value and Prices*. New Haven, CT: Yale University Press.

Genesove D, Mayer C (2001) Loss Aversion and Seller Behavior: Evidence from the Housing Market. *The Quarterly Journal of Economics* 116:1233–1260.

Genevsky A, Yoon C, Knutson B (2017) When Brain Beats Behavior: Neuroforecasting Crowdfunding Outcomes. *The Journal of Neuroscience* 37:8625–8634.

Glimcher PW, Fehr E (2014) Introduction: A Brief History of Neuroeconomics. In: *Neuroeconomics*, pp xvii–xxviii. Elsevier. Available at: https://linkinghub.elsevier.com/retrieve/pii/B9780124160088000358 [Accessed June 30, 2024].

Glimcher PW, Rustichini A (2004) Neuroeconomics: The Consilience of Brain and Decision. *Science* 306:447–452.

Glimcher PW, Tymula AA (2023) Expected Subjective Value Theory (ESVT): A Representation of Decision Under Risk and Certainty. *Journal of Economic Behavior & Organization* 207:110–128.

Gul F, Pesendorfer W (2008) The Case for Mindless Economics. In: *The Foundations of Positive and Normative Economics* (Caplin A, Schotter A, eds), pp 3–42. Oxford University Press. Available at: http://www.oxfordscholarship.com/view/10.1093/acprof:oso/9780195328318.001.0001/acprof-9780195328318 [Accessed June 30, 2024].

Hardie B, Johnson EJ, Fader P (1993) *Modeling Loss Aversion and Reference Dependence Effects on Brand Choice*. Available at: https://papers.ssrn.com/abstract=1424038 [Accessed June 30, 2024].

Hartman RS, Doane MJ, Woo C-K (1991) Consumer Rationality and the Status Quo. *The Quarterly Journal of Economics* 106:141–162.

Holzmeister F, Huber J, Kirchler M, Schwaiger R (2022) Nudging Debtors to Pay Their Debt: Two Randomized Controlled Trials. *Journal of Economic Behavior & Organization* 198:535–551.

Johnson EJ, Goldstein D (2003) Do Defaults Save Lives? *Science* 302:1338–1339.

Kable JW, Glimcher PW (2007) The Neural Correlates of Subjective Value During Intertemporal Choice. *Nature Neuroscience* 10:1625–1633.

Kable JW, Glimcher PW (2010) An "As Soon as Possible" Effect in Human Intertemporal Decision Making: Behavioral Evidence and Neural Mechanisms. *Journal of Neurophysiology* 103:2513–2531.

Kahneman D, Knetsch JL, Thaler RH (1990) Experimental Tests of the Endowment Effect and the Coase Theorem. *Journal of Political Economy* 98:1325–1348.

Kahneman D, Knetsch JL, Thaler RH (1991) Anomalies: The Endowment Effect, Loss Aversion, and Status Quo Bias. *Journal of Economic Perspectives* 5:193–206.

Kahneman D, Tversky A (1979) Prospect Theory: An Analysis of Decision Under Risk. *Econometrica* 47:263–292.

Knutson B, Genevsky A (2018) Neuroforecasting Aggregate Choice. *Current Directions in Psychological Science* 27:110–115.

Köszegi B, Rabin M (2006) A Model of Reference-Dependent Preferences. *The Quarterly Journal of Economics* 121:1133–1165.

Koszegi B, Rabin M (2007) Reference-Dependent Risk Attitudes. *American Economic Review* 97:1047–1073.

Laibson D (1997) Golden Eggs and Hyperbolic Discounting. *The Quarterly Journal of Economics* 112:443–478.

Lempert KM, Glimcher PW, Phelps EA (2015) Emotional Arousal and Discount Rate in Intertemporal Choice Are Reference Dependent. *Journal of Experimental Psychology: General* 144.

Lempert KM, Phelps EA (2013) Neuroeconomics of Emotion and Decision Making. In: *Neuroeconomics*, pp 219–236. Elsevier. https://doi.org/10.1016/B978-0-12-416008-8.00012-7.

Lerner JS, Li Y, Valdesolo P, Kassam K (2014) Emotion and Decision Making. *Annual Review of Psychology* 66:799–823.

Lewis M (2017) *The Undoing Project: A Friendship that Changed Our Minds*, 1st ed. New York: W.W. Norton & Company.

Mankiw NG (2000) *Principles of Economics*, 2nd ed. Fort Worth, TX: Hartcourt College Publishers.

Markowitz H (1952) The Utility of Wealth. *Journal of Political Economy* 60:151–158.

Marshall A (1997) *Principles of Economics*. Amherst, NY: Prometheus Books.

McClure SM, Laibson DI, Loewenstein G, Cohen JD (2004) Separate Neural Systems Value Immediate and Delayed Monetary Rewards. *Science* 306:503–507.

Milkman KL, Beshears J, Choi JJ, Laibson D, Madrian BC (2011) Using Implementation Intentions Prompts to Enhance Influenza Vaccination Rates. *Proceedings of the National Academy of Sciences of United States of America* 108:10415–10420.

Ogawa S, Tank DW, Menon R, Ellermann JM, Kim SG, Merkle H, Ugurbil K (1992) Intrinsic Signal Changes Accompanying Sensory Stimulation: Functional Brain Mapping with Magnetic Resonance Imaging. *Proceedings of the National Academy of Sciences* 89:5951–5955.

Phelps EA, Lempert KM, Sokol-Hessner P (2014) Emotion and Decision Making: Multiple Modulatory Neural Circuits. *Annual Reviews of Neuroscience* 37:263–290.

Platt ML, Glimcher PW (1999) Neural Correlates of Decision Variables in Parietal Cortex. *Nature* 400:233–238.

Samuelson PA (1937) A Note on Measurement of Utility. *The Review of Economic Studies* 4:155.

Samuelson W, Zeckhauser R (1988) Status Quo Bias in Decision Making. *Journal of Risk and Uncertainty* 1:7–59.

Shleifer A (2000) *Inefficient Markets: An Introduction to Behavioral Finance*, 1st ed. Oxford: Oxford University Press. Available at: https://academic.oup.com/book/27761 [Accessed June 30, 2024].

Sokol-Hessner P, Camerer CF, Phelps EA (2013) Emotion Regulation Reduces Loss Aversion and Decreases Amygdala Responses to Losses. *Social Cognitive and Affective Neuroscience* 8:341–350.

Sokol-Hessner P, Hsu M, Curley NG, Delgado MR, Camerer CF, Phelps EA (2009) Thinking Like a Trader Selectively Reduces Individuals' Loss Aversion. *Proceedings of the National Academy of Sciences of the United States of America* 106:5035–5040.

Sokol-Hessner P, Lackovic SF, Tobe RH, Camerer CF, Leventhal BL, Phelps EA (2015) Determinants of Propranolol's Selective Effect on Loss Aversion. *Psychological Science* 26:1123–1130.

Thaler R (1980) Toward a Positive Theory of Consumer Choice. *Journal of Economic Behavior & Organization* 1:39–60.

Thaler RH, Benartzi S (2004) Save More TomorrowTM: Using Behavioral Economics to Increase Employee Saving. *Journal of Political Economy* 112:S164–S187.

Thaler RH, Sunstein CR (2003) Libertarian Paternalism. *The American Economic Review* 67:420.

Thaler RH, Sunstein CR (2008) *Nudge: Improving Decisions About Health, Wealth, and Happiness*. New Haven: Yale University Press.

Tom SM, Fox CR, Trepel C, Poldrack RA (2007) The Neural Basis of Loss Aversion in Decision-Making Under Risk. *Science* 315:515–518.

Tversky A, Kahneman D (1981) The Framing of Decisions and the Psychology of Choice. *Science* 211:453–458.

Von Neumann J, Morgenstern O (1944) *Theory of Games and Economic Behavior.* Princeton, NJ: Princeton University Press.

3

Neuroscience and the Free Will Debate

Natalia Peraza and John R. Monterosso

Could neuroscience settle the longstanding philosophical debate over whether people possess free will? That is a lot to ask, but we will argue here that neuroscience can provide a basis for evaluating free will, at least in one sense that the term is used. To develop this argument, we need to first set the table by clarifying two very different notions of what it is to have free will. The starting point for both of these is the special way we ordinarily make sense of one another's behavior, which is qualitatively different from the way we explain the rest of the world.

3.1 The Physical World vs. Mind World

What makes a lightning bolt strike? Or mud slide down a canyon after a rainstorm? Or a cresting wave break? These are events in the physical world, and in modern science-grounded cultures, we primarily explain them by identifying *prior conditions and laws* from which the events followed. An explanation in this *physical world framework* might be superficial and vague or thorough and precise, but in either case, the content of the explanation is made up of elements present prior to the event being explained. The future is irrelevant to physical world framework explanations. For example, within this framework, it would not make sense to explain a lightning bolt by noting the fire that came *after* it struck a house. It would, however, make sense to explain it by citing the charge differential between the lower and upper portions of clouds just *prior to* the lightning. Aristotle called these type of explanations, on which we rely heavily (though not exclusively) in science, "efficient causes" (Falcon, 2006). An efficient cause of an event is an antecedent that set the event in motion.

Within the physical world framework (at least at the supra-atomic level), events follow inevitably from causes. The side of a functioning balance scale with more weight does not just *usually* go down, it *always* goes down. Of course, our physical world framework understanding is often incomplete, and so our ability to predict events is limited. After a long drought followed by a heavy rain, it is a good bet there will be a mudslide *somewhere*, but the exact time, place, and severity of a slide cannot be predicted. This limitation is not because the sliding of mud is any less lawfully determined than the tipping

DOI: 10.1201/9781032717791-3

of a balance scale. But mudslides involve more complex systems over large areas, and so predicting a specific occurrence is not currently possible. Nevertheless, just as our ability to predict the movements of hurricanes has steadily improved, year by year, with advances in technology and analytic methods (https://ourworldindata.org/weather-forecasts), we can plausibly imagine a future in which mudslides become predictable with far greater specificity. In the physical world framework, everything happens in a deterministic causal chain, and the idea of "free will" simply does not apply.

By contrast, when answering "Why?" questions about other people, we ordinarily engage an entirely different explanatory framework, which we will call the *mind framework*. In the mind framework, we explain actions by inferring the individual's intentions and sometimes the motivations and beliefs that supported the intentions. Is the person up ahead waving her hand trying to say hello to you or someone else? Or maybe she is just trying to swat away a bee. In either case, she is *trying* to do something. She has an intention. Intentions imply the pursuit of a goal state – a *future* that is being sought. So, unlike the physical world framework, where explanations look exclusively back in time, within the mind framework, we partly answer "why?" questions by identifying a future state being pursued (or avoided). Put differently, within the mind framework, we can explain events by identifying their *purpose*. The Greek word for purpose is "telos," and so Aristotle called these explanations, in which the observer understands an action by identifying the intention, "teleological." If we use the mind framework instead of the physical world framework to explain a lightning bolt (maybe we think it is an act of an angry god), then the housefire that would predictably follow is no longer irrelevant to explaining the lightning. If the lightning bolt was the act of Zeus's mind, then the fact that it was going to result in the destruction of a home very well could be a key component of our understanding of why lightning occurred. This is an important difference – anticipated future states have no place in explanations based in the physical world framework, but they are often essential to mind framework explanations.

Somewhat complicating matters, there are times in which teleological explanations are given, even when we don't think the behaving entity has a mind. For example, in response to the question, "Why does the tobacco plant use energy to synthesize nicotine?", a reasonable answer is that it does so for the purpose of killing off pests. The phrase, "for the purpose of," or something along those lines, is the giveaway that this is a teleological explanation. But in such a case, the intention is an "as if" state that emerges because the trial and error of natural selection results in outcomes that approximate what intention might produce (what Richard Dawkins had in mind when he referred to natural selection as a "blind watch maker" (1996)). But we don't think the plant *actually* holds the intention to kill bugs in the way that we think an on-the-job exterminator holds that intention. In ordinary person perception, intentions are actual states, not just ways to describe behavior. That is made clear by the fact that we think we can be wrong about intentions. Maybe you thought a person's teasing was intended to hurt your feelings but come to learn that it was actually playful, intended to bring you closer. The fact that we can be wrong about an intention conveys that we view it as an actual state (and not just an "as if" short-hand). One can even imagine, in the future, using a brain scan to confirm a person's actual intention, whereas it of course would not make sense to use a brain scan to measure the plant's true intention, or, at least, it would not make sense unless we thought plants had minds.

Occasionally there are high-stakes interpersonal situations in which intentions are ambiguous ("Is he reaching for his wallet or his gun?"), but more often intentions are apparent ("He is reaching for his wallet to pay the cashier."). We make sense of one another's behavior by effortlessly and automatically inferring intentions, as well as associated beliefs and motivations. We don't need to be taught the mind framework, because our social brains evolved to understand each other this way (Dunbar, 1998; Dunbar & Shultz, 2007). Indeed, our disposition to use the mind framework is so robust that it can be triggered in situations in which we don't really think a mind framework is appropriate, as demonstrated to comic effect by the clever IKEA lamp commercials of Spike Jonez (https://www.youtube.com/watch?v=ecTUnfHyj8k, and see also the classic 1944 animation of social psychologists Heider and Simmel, https://www.youtube.com/watch?v=VTNmLt7QX8E). The two broad conceptions of free will, which we turn to next, are both integrally linked with this mind framework.

3.2 Libertarian Free Will

Imagine you are at a dinner party and your host asks you whether he can get you a second scoop of ice cream. The first scoop was delicious, and the thought of more pulls you in. But you also think about your interest in losing a few pounds. For a moment you pause, stuck between two possibilities that each have some appeal, and then you decide it is just too good to resist and so you go for seconds. Given everything about you in that moment of choice – your experiences, your personality, your genetics, your brain state down to the molecular level – was that choice inevitable? Or could you have done otherwise? Put differently, if the universe could somehow be rewound and set in motion again just before your choice, is it possible that you could resist the temptation on the rerun? If you think the answer is yes, then you believe in what philosophers refer to as "libertarian free will." The libertarian free will position ascribes at least some human behavior to what philosophers call *causa sui* (Latin for "self-caused") (Palmer, 2014), since an act of free will in this sense is not just the consequences of antecedents in a deterministic causal chain. It is the initiation of something new.

To many, the libertarian free will position seems intuitive within the mind framework (Ekstrom, 2005; Kane, 2018). While the causes of the physical world framework are expected to function deterministically (e.g., once a ball has been hit, precisely where it goes is 100% a function of the physical forces that act on it), we tend to look at other people's actions, or our own actions, differently. The ball goes where it goes, and it cannot be otherwise, but the behavior of the batter may feel different. How hard he runs, and indeed whether he runs at all, seems to allow indeterminacy. The ball must go where it has been sent, but not the batter. When we ask a lecture hall full of students to raise a hand if they believe they have free will in the "could have done otherwise" libertarian sense, most hands go up.

3.3 Neuroscience and Libertarian Free Will

How does the idea of libertarian free will square with neuroscience? We will argue that the answer is: not very well! Some have gone so far as to conclude that neuroscience actually empirically disproves libertarian free will, based on a series of studies going back over 40 years. In the earliest of these studies, Benjamin Libet and colleagues (1983) utilized a device called an electroencephalogram (EEG), which uses electrodes placed on the scalp to measure electrical activity produced by populations of neurons. Participants were instructed that they should, from time to time, make a hand movement. Since the researchers wanted to create a situation in which the participant felt maximally free to choose, it was left entirely up to them to decide when to make each movement, and each participant had one additional job – she was asked to notice where a running timer was at the moment she *decided* to move and to report that time after each movement. So, for example, a participant might report that she formed the intention to make her first movement at the moment the counter was at 18 seconds. The researchers also recorded when each movement *actually* occurred, based on direct recording of electrical activity at the wrist. So at the end of each experimental session, the researchers had three kinds of data: 1) a series of times in which the participant reported forming an intention to move, 2) a series of times in which a movement actually occurred, and 3) EEG data indicating brain activity throughout the session.

Libet and colleagues were interested in using their data to address whether conscious decisions to move (intentions) occurred too late to be the actual drivers of movement. To this end, the researchers capitalized on the well-established finding (Gilden et al., 1966) that there is a ramping up of EEG signal emanating from the supplementary motor region (just anterior to the primary motor cortex) prior to the execution of self-directed movements. This pre-movement EEG signal is commonly referred to as a "readiness-potential" (RP). Libet and colleagues reasoned that if the intention truly initiates the movement, then the intention should come first, followed by the preparatory RP, and then by the actual movement. But that's not what happened. On-average, the brain's RP indicating that a hand movement was about to occur happened 8 tenths of a second before the intention (as reported by participants). The authors concluded, "the brain evidently 'decides' to initiate or, at the least, prepares to initiate the act at a time before there is any reportable subjective awareness that such a decision has taken place" (Libet et al., 1983, p. 640). If the gears of the nervous system that drive movement are already in motion when intentions to move occur, it suggests that intentions don't really *cause* the movement and that the feeling to the contrary is an illusion of will, a "willusion" (Nahmias, 2011). Subsequent studies reported similar results using different neural measurements, including functional magnetic resonance imaging (Soon et al., 2008) and even single cell recording (Fried et al., 2011).

These are interesting experiments, and they add to a great deal of research in the behavioral sciences that suggests our own reports about intentions are often wrong (Gazzaniga,

1967; Haidt, 2001; Nisbett & Wilson, 1977). But the Libet studies do not settle the free will debate. One issue is that the previous strong inference depends on participants' accuracy in reporting where a running timer was when their intention arose, and participants don't appear to be very good at this. Seemingly superficial changes in the procedure have been shown to significantly affect the experimental results (Bredikhin et al., 2023; Sanford et al., 2020). Moreover, even if we grant that there is some ability to predict action before an intention, as claimed by Libet, that does not mean that intention was *settled* at the point it can be predicted (Nahmias, 2014). Perhaps the initial ramping up of the readiness potential is indexing a point in which an eventual intention is becoming more likely but not yet certain. If so, the subsequent moment of intention could still mark the point at which the behavior was truly set in motion, which is to say, temporal precedence of the RP over the intention does not imply the intention has no force. The sensation of an itch regularly precedes the intention to scratch, but it would be silly to suggest that the decision to scratch was illusory because it was predictable given the itch (or the itch's neural correlates). Finally, those who hold the libertarian free will position need not believe that all or even most human behavior has the "causa sui"/"could-have-done-otherwise" quality. They could maintain that acts of free will only occur on occasions where there is something so important to the person that she overrides the normal flow of cause and effect, and so even if neuroscientists could convincingly demonstrate its absence in controlled settings, it would be difficult to rule it out altogether on that basis.

And yet, while we don't think neuroscience has provided an empirical basis for ruling out libertarian free will, commitment to neuroscience, or indeed to any science of human behavior, favors rejecting it, at least so long as one is engaged in the practice of science. Libertarian free will implies dualism. It implies a force that stands apart from laws that govern the physical world but that nevertheless interacts with it. Whatever else you might think people are, we are bodies in the physical world. We are (mainly) oxygen, carbon, hydrogen, nitrogen, calcium, and phosphorus. And all those elements in a human body could be analyzed from the physical world framework, perhaps by some futuristic machine with unimaginable capacity for computation. In any moment, including the moment we contemplated whether to have seconds of dessert, the countless molecules of a body are in a particular state. And given any state, there is a *next state* that would follow by the laws of physics. But the libertarian free will position implies that the next state designated by physical law will not happen if an act of human *will* intervenes. This would set beings with libertarian free will apart from the rest of the universe, which is what Nietzsche (1886) had in mind when he referred to libertarian free will as "nonsense" and the product of man's "extravagant pride." Libertarian free will implies that behavioral science is not just currently incomplete; rather it can *never* be complete. If it were true that given the exact same state, a person nevertheless *could have done otherwise*, then it follows her behavior is inherently *unpredictable*. Libertarian free will cannot be incorporated into neuroscience because the idea stands in conflict with the entirety of the project. Neuroscience ends wherever one thinks libertarian free will begins.

3.4 Neuroscience and Compatibilist Free Will

Does belief in neuroscience's potential to explain behavior require rejecting the concept of free will entirely? Not necessarily. There is an alternative (or, really, family of alternatives) to the libertarian conception of free will that does not contradict neuroscience. Like libertarian free will, this conception is grounded in the mind framework. But unlike libertarian free will, it is *compatible* with conceiving humans as (also) fully part of the physical world and as fully subject to deterministic laws. This is a version of free will that does not require that people "could have done otherwise" in the sense described previously. Because of its compatibility with determinism, this alternative is referred to as *compatibilism*. In a recent survey of 648 philosophers, many more endorsed agreement with the idea of compatibilist free will than libertarian free will, by a margin greater than 3 to 1 (Bourget et al., 2023). Of course, popularity of a view does not guarantee it is correct, but it is nevertheless worth noting, since this is a group that has given the matter considerable thought and debate. And in that context, the idea of compatibilist free will predominates.

Broadly, compatibilists view "free will" as present when an act has the right sort of connection to the individual's mental state. Various proposals have been made regarding what exactly that relationship must be. Thomas Hobbes (1676) proposed a broad definition, where freedom is simply the absence of external constraints on fulfilling one's desires. So, simply, if you can do what you want to do, you are free. Alternatively, Harry Frankfurt (1971) argued for a hierarchy of desires, prioritizing "higher-order desires" (e.g., the desire to quit a harmful addiction) over immediate "first-order desires" (e.g., the desire to get high). In Frankfurt's more restrictive account, acting on a first-order desire could be considered unfree if it were contrary to one's higher-order desire (1971) (for discussion of grounding higher-order desires in the hierarchical organization of the brain, see Felsen and Reiner (2011)). Other compatibilists have argued that to be free, behavior must be modifiable based on new reasons (Dennett, 1984). Alternatively, Peter Strawson (1974) grounds free will in innate "reactive attitudes" of others, such as resentment or gratitude, which presuppose that the actor is responsible for their choices (Strawson, 1974).

Fleshing out these compatibilist proposals is beyond the scope of this chapter. For our purposes, it is sufficient to adopt the simple compatibilist criterion that Hobbes proposed – for an act to be free, it must be the case that *you did what you wanted to do*. This simple formulation can be used to clarify the ways in which neuroscience could adjudicate when the will is free. We will specifically focus on two words in that formulation "want" (was the action generated in connection with a desire?) and "you" (was the source of the desire really "you"?). Our goal is to describe how neuroscience, and especially the sub-area of neuroeconomics, might be used to adjudicate whether an act is free in the compatibilist sense. Throughout, we will use the phenomenon of addiction as an illustrative example.

3.5 Neuroscience and Habit Control (Doing without Wanting)

One way that an action can fail the "you did what you wanted to do" test is if the *want* element is missing. Charles Sherrington famously inferred the basic features of synaptic communication through the study of a simple leg extension reflex. The triggering stimulus of a reflex does not depend on wanting the reflex to occur. Regardless of what a person does or does not want, their pupils will constrict in bright light, and their sweat glands will secrete when the temperature is high. Indeed, many reflexes will occur for some time after decapitation. Since reflexes are not a product of what the individual wants, they are not free in the compatibilist sense.

A more interesting case in which behavior *may* fail the "wanting" criterion involves the formation of strong habits. Intuitively, we may sometimes experience ourselves doing something that we do not want to do *out of habit*. Suppose most every day I drive a particular route to work. But today, I happen to be on an errand that requires that I start driving my usual path but then take a different turn midway into it. If I get lost in thought or in the music I am listening to, I might find myself staying on the usual path to work beyond the point where I should have diverged. At no time did I *want* to go to work, so this habit-controlled behavior fails the "wanting" criterion and is not an act of free will in the compatibilist sense.

In behavioral neuroscience, a clever procedure can be used to distinguish whether a particular behavior is controlled by 1) habit or 2) wanting (with the latter referred to as "goal-directed"). In the procedure, an animal is trained to make some arbitrary response, such as pressing a lever to obtain a food reward. After sufficient training, the animal reliably presses the lever whenever it is given access to it. Is the animal pressing the lever because it wants the reward? Or is the behavior so *habitual* that it has become as automatic as a reflex, perhaps bypassing wanting entirely? To test this, after learning takes place, the researcher removes the animal from the cage with the lever and then does something that changes the value of the reward. For example, if the reward was a bit of dried fruit, the researcher can give the animal a bit of the same flavored dried fruit followed by lithium chloride injection. The lithium chloride will make the animal nauseated, and that pairing will establish a "taste aversion" such that the rat will come to dislike the taste, often for the rest of its life. The critical question is: What will happen if that animal is then returned to the cage in which it was trained to press a lever, with presses rewarded with dried fruit? Will it press the lever, even though doing so will produce a food that it will not eat? If it does, then the behavior is not based on wanting, or at least not on wanting the food.

So what actually happens in this procedure? That depends on the length of the training. Generally, if the training history (the number of learning trials before the nasty intervention to produce an aversion to the reward) was relatively short, the animal will not press the lever. Evidently, the animal's response in this case is controlled by its valuation of what it knows will follow. And since what will follow is something it does not want, the animal does not press the lever. However, if there has been an extensive learning history, with hundreds or even thousands of lever presses rewarded with the food prior to creation of the taste, that makes a big difference. Now the animal will typically go right ahead and press the lever, even though the lever leads to delivery of a food it will not eat. Something evidently changes with excessive learning such that control of the behavior is no longer linked to the desirability of the reinforcer used during training (Dickinson, 1985). The behavior has become a "habit" (Robbins & Costa, 2017). Using this "goal-devaluation" paradigm, along with other experimental methods, neuroscientists have made great progress delineating and dissociating the neural circuits that control goal-directed (wanting-based) vs. habit-driven responses (Belin et al., 2009). In general terms, the ventral striatal–cortical loops that control behavior early in learning (when behavior is goal-directed) give way to dorsolateral striatal–cortical loops as control shifts. This ventral to dorsal control shift results in faster and more efficient action control but at the expense of flexibility to changes in conditions, including changes to whether the outcome is still desirable. And it is this shift that makes the behavior a "habit."

Since habits don't depend on wanting, is behavior controlled by habit unfree in the compatibilist sense? Put differently, does it fail the criterion, "Did you do what you *wanted* to do?" Certainly not always. Most of our habits are consistent with our goals (James, 1887). Indeed, in many cases, we cultivate the habits we want. And even when habits clash with what we want, they are often easy to override. My drive-to-work habit might have led me to a wrong turn when I was supposed to drive to the market, but the moment I become consciously aware of my mistake, I have no problem course-correcting. But some have argued habits are sometimes so strong that they overpower what the individual wants. In particular, it has been suggested that at least some addictions can be explained as habits that persist even when the individual wants to stop. Most prominently, this argument has been advanced by neuroscientists Barry Everitt and Trevor Robins. In their theory of addiction, they emphasize the role of habits, suggesting that the drug use habit takes on a persistent quality that differentiates it from goal-directed behavior. They explicitly object to the characterization of the individual's state as one of subjective wanting (as emphasized in incentive-sensitization models), arguing instead that it "corresponds more obviously to the subjective state of 'must do!' – although this subjective response could arise post hoc as a rationalization of the 'out-of control' habitual behavior rather than being the driving influence" (Everitt & Robbins, 2005, pg 1485). The suggestion here is that the reports of drug *wanting* are misleading, as the actual behavior is under habit control (like the rat pressing a lever for food it does not value). There is reason to be skeptical of this as a full explanation of addiction (Heyman, 2009; Lewis, 2017). But the key point here is that advances in neuroscience provide a potential basis for assessing when the wanting piece of the "you did what you wanted to do" criterion is missing and so a basis for concluding a behavior unfree. However, even presuming a futuristic technology that decodes intention from brain activity (Andersen & Ross, 2019), it is important to keep in mind that multiple networks are involved in even the simplest behaviors, so attributing a behavior exclusively to the habit system is at best an approximation.

3.6 Neuroscience and Multi-Agent Mind Proposals (Is It Always *You* Carrying Out the Desired Behavior?)

Within behavioral science and philosophy, the term "agent" refers to an entity that acts with intention (Anscombe, 1956; Davidson, 2015). It can be used to describe a person or a group that acts as a collective, such as a company. Most importantly for the present discussion, an "agent" can also be used to characterize multiple systems within a single person. In such cases, the person is conceived as containing multiple agents within their mind. The history of such "multi-agent mind" theories long predates neuroscience. Plato conceived of the individual as containing three distinct parts, each with its own values. He analogized the three parts to two horses and one charioteer at the reins. One horse was moved by basic appetites and the other by noble desires, and the charioteer was governed by reason and wisdom. Because they want different things, Plato's three parts are treated as distinct agents (Santas, 2006). Similar conceptualizations can be found in religious texts. For example, in the New Testament, Romans 7 (Trobisch, 2000), the individual is conceived as a body governed by sin that competes with the spiritual self. In words attributed to Apostle Paul, "I do not understand what I do. For what I want to do I do not do, but what I hate I do. And if I do what I do not want to do, I agree that the law is good. As it is, it is no longer I myself who do it, but it is sin living in me." In Paul's conception, there is a true self, which is identified with the individual's acceptance of spiritual values, but there is a separate force of sin which moves the body. And the text here suggests that Paul viewed his own sinful behavior as a loss of freedom in the compatibilist sense, since it was not his true "I." Multi-agent models may have important implications for compatibilist free will because the agents may not be considered equally *you*.

Is there any support to be found in modern psychology and neuroscience for multi-agent mind theories? Before considering any specific multi-agent proposals, it is useful to introduce an extreme case which demonstrates that the neural basis of decision-making is more fragmented than introspection might suggest. In the 1960s, Michael Gazzaniga and Roger Sperry conducted groundbreaking research on patients who had undergone corpus callosotomy, a surgical procedure that severs the corpus callosum connecting the brain's two hemispheres (Gazzaniga, 1967). This surgery, typically performed to alleviate severe epilepsy, provided a unique opportunity to study how the isolated cortical hemispheres function independently, since the surgery eliminated inter-hemispheric cortical communication. The behavior of these patients highlighted the independent capabilities of each hemisphere, occasionally leading to conflicting actions – such as one hand buttoning a shirt while the other unbuttoned it. In such a case, would it make sense to consider the hemispheres as two selves? And if so, how much communication between them would it take before the two hemispheres could be thought of as one self?

Perhaps the most intriguing finding to come out of Sperry and Gazzaniga's work was what happened when the "split-brain" patients were asked about intentional behavior that originated in the right hemisphere (Gazzaniga, 2014). For example, suppose a patient had just complied with the command that she get a soda, which had been presented exclusively to her right hemisphere. Now she is asked, "Why did you get a soda?" Since speech is strongly dominated by the left hemisphere, the left hemisphere will control the answer she gives. But since her corpus collosum has been severed, the left hemisphere has no access to the fact that she had been told to get the soda. So, what does the patient say? Does she say something like, "I have no idea, I was just sitting here and suddenly I found myself getting up and I don't even know why?" The interesting and well-replicated finding is that she does not. Indeed, she does not appear befuddled by her behavior at all. Instead, she offers an answer that would seem plausible (e.g., "I was thirsty") were it not for the fact that the experimenter knows something that the left hemisphere doesn't know.

You might chalk this up to the patient trying to get out of an awkward spot, but extensive work with these patients suggests that the reality is far more interesting. By all indications, *the patient believes her answers*. It is this false belief of accuracy that makes this response a "confabulation" rather than a lie. Work in experimental (Johansson et al., 2005; Johnson & Raye, 1998) and social psychology (Nisbett & Wilson, 1977) suggests that confabulation regarding the reasons for behavior is not unique to split-brain patients. Michael Gazzaniga has argued convincingly that the confabulation of the split-brain patient, while striking, is not qualitatively different from how we normally understand our own behavior. Perhaps because we are highly social creatures who need to justify what we do, human brains evolved a system to introspect and infer reasons for our actions, which Gazzaniga calls our "interpreter module." But that system is far from omniscient. We do not have awareness of the great majority of the information and processes that drive our own decisions, and so our explanation of why we do what we do is on shaky ground – an interpretation based on limited information. He writes,

> The interpreter receives the results of the computations of a multitude of modules. It does not receive the information that there are multitudes of modules. It does not receive the information about how the modules work. It does not receive the information that there is a pattern-recognition system in the right hemisphere. The interpreter is a module that explains events from the information it does receive.
>
> **(Gazzaniga quoted in Street, 2017)**

If Gazzaniga is right, that inner voice in your head that feels like the CEO is more analogous to your brain's press secretary (Haidt, 2006). Like a press secretary, the interpreter module has the job of explaining and justifying what has been done but is not itself in a position to make something happen and is often not even privy to the factors that actually drove action. The demotion of the inner voice from CEO to press secretary has important implications for evaluating free will in the compatibilist sense. If the "you" suggested by introspection is really just the voice of an interpreter module, there is no longer an easy basis to say all intentional behavior came from "you." Instead, absent a captain at the helm, we may look at competing systems and treat only one of those systems as the true you. In such case, some motivated behavior could be considered unfree, because the source was not "you." Whether the data support this is an empirical question.

3.7 Specific Multi-Agent Mind Accounts

There is a 1987 commercial for a doughnut shop chain (https://www.youtube.com/watch?v=sjqUMOdgH5E) that opens with Fred the baker in a dark bedroom. It is clearly not yet morning, but duty calls – it is "time to make the doughnuts." A devil appears on one shoulder, encouraging Fred to stay in bed and get a good night's sleep for once. An angel appears on the other shoulder reminding Fred that people will want fresh doughnuts. We next see Fred hard at work, as the narrator comments, "It takes a lot of willpower to make Dunkin' Donuts fresh day and night." Within psychology, several attempts have been made to explain the struggle between doing what is most immediately gratifying vs. what we know we ought to do as the product of multiple competing agents within us, like Fred's shoulder spirits. Freud conceived of the individual as divided into three parts, each with its own values: 1) the id, the seat of drive and emotions, governed by the pleasure principle; 2) the ego, which considers longer-term interests and is governed by the reality principle; and 3) the superego, which is based on the internalization of rules of society or of parents and is governed by the morality principle. Several more modern proposals distinguish between two systems (System 1 and System 2 models) that are sometimes in competition. System 1 is evolutionarily older, operates rapidly without conscious effort, and is driven by immediate stimuli. System 2 is evolutionarily newer; operates slowly, requiring focused attention; and employs abstract thinking, enabling planning and future-oriented actions (Kahneman, 2011; Street, 2017). Proposals in this family include Thaler and Shefrin's synchronic multiple-self model, which proposed that individuals possess "two sets of coexisting and mutually inconsistent preferences: one focused on the long term and the other on the short term" (1992). Similarly, Metcalfe and Mischel's exploration of the "dynamics of willpower" suggests the existence of multiple agents within an individual: one represented by a cool, cognitive "know" system and the other by a hot, emotional "go" system (1999).

A commonly held view is that the functioning of the lateral prefrontal cortex instantiates cool System 2 motives, while sub-cortical circuits (and in some models midline structures of the prefrontal cortex) instantiate hot System 1 motives. For instance, Shefrin and Thaler argued that their theory was

> roughly consistent with the scientific literature on brain function. . . . The planner in our model represents the prefrontal cortex. The prefrontal cortex continually interfaces with the limbic system, which is responsible for the generation of emotions. The doer in our model represents the limbic system.
>
> **(Shefrin & Thaler, 1992)**

It is possible to treat the agents in a multi-agent mind model as all part of a single "you," analogous to different states making up a unitary federation. But is that how people think about identity? Or, when answering that key free will question, "Did you do what you wanted to do?", is one system viewed as more central to "you"? If the latter possibility is true, then the tools of neuroscience could be used to determine whether a behavior is free by identifying the relative role each system played in the behavior.

3.8 Neuroeconomics and the Search for Multi-Agent Mind Models

The idea of competing systems has been explored extensively in neuroimaging work on "delay discounting" tasks over the past two decades. These tasks require participants to make a series of choices between a reward that is more immediate vs. one that is greater but is more delayed. If there are indeed separate motivational systems that are differentially focused on the short and long term, delay discounting tasks should provide a systematic way to examine how competition between these systems unfolds. The first study to pair neuroimaging with intertemporal choice was a neuroeconomics "dream team" collaboration between psychologists Sam McClure and Jonathan Cohen and behavioral economists David Laibson and George Loewenstein. In their study (which has been cited more than 4000 times, according to Google Scholar), the rewards at stake for participants were gift certificates for Amazon.com ranging from $5 to $40. The smaller sooner option varied in delay between the same day ("today") and 4-week delay; the larger later option was always available either 2 or 4 weeks after the smaller sooner option. The team used a multi-agent economic model designed to capture the influence of a hypothesized impatient System 1 that only valued immediate reward and a more rational System 2 that valued the long term along with the short term. Their model used the Greek letter b (beta) to denote System 1 and d (delta) for System 2 (McClure et al., 2004).

McClure and colleagues used their beta-delta multi-agent model to first identify possible neural substrates of distinct systems and then to investigate whether there was evidence that these systems competed, like the spirits on the Dunkin Donuts baker's shoulders. The results of their initial study were supportive of this possibility. Specifically, a primarily limbic/paralimbic ("beta") network of brain regions appeared to become active only in connection with alternatives that included an immediate reward option. Another ("delta") network that included the lateral prefrontal cortex responded to all rewards, even when they were delayed. What's more, the activity in these two networks during decision-making suggested system competition, with the delta network activity stronger on trials in which participants were willing to wait for the larger amount and beta network activity higher when participants chose the immediate money. The authors interpreted their findings as indicative of an internal battle of systems that resemble the present-focused Grasshopper and future-oriented Ant of Aesop's cautionary fable. According to their model, the selected behavior often reflects the outcome of a

> competition between lower level, automatic processes that may reflect evolutionary adaptations to particular environments, and the more recently evolved, uniquely human capacity for abstract, domain-general reasoning and future planning . . . preferences seem to reflect a competition between the impetuous limbic grasshopper and the provident prefrontal ant within each of us.
>
> **(McClure et al., 2004, pg 507)**

In addition to potentially informing questions of free will, findings like the McClure team's had important implications not only for understanding of willpower struggle but potentially for interventions as well. For example, individuals struggling to do what is in their long-term interest (e.g., quit smoking, eat healthy) could be assessed for the contribution of weak System 2 vs. strong System 1, with potential implications for appropriate interventions.

However, subsequent work has led to a more complicated picture than suggested previously (Kable & Glimcher, 2007, 2010). Kable and Glimcher (2007) observed that activation within the network McClure and colleagues identified as "beta regions" was not differentially sensitive to immediacy but rather tracked overall value. And our own team's subsequent research on value tracking outside of a decision context indicated that hyper-responsivity to immediate rewards included increased signal in the dorsolateral prefrontal cortex (delta regions), along with the expected hyperactivity in the limbic and paralimbic system (Luo et al., 2011). Using a network approach, Elton and colleagues (2017) reported that individuals more willing to wait for larger rewards exhibited more activity in a network that included lateral regions of the prefrontal cortex but *also the striatum* (Elton et al., 2017). Interestingly, that study reported evidence of network competition, but they observed a temporal lobe network (rather than limbic/paralimbic regions) to be associated with high preference for immediate reward.

An important critique to system competition perspectives on self-control relates to the independence of the systems. There is considerable evidence that the lateral prefrontal cortex subserves long-term interests by modifying (rather than competing with) signaling in limbic regions (Hare et al., 2009). In other words, while delay of gratification appears to depend on capacities supported by the prefrontal cortex, those capacities affect intertemporal choice through *mediation of* (rather than *competition with*) older cortical and subcortical structures central to reward and motivation (Monterosso & Luo, 2010).

3.9 Addiction, Multi-Agent Models, and Loss of Free Will

The previous evidence implicating lateral sectors of the prefrontal cortex as important for the human capacity to pursue temporally distant rewards may be important in understanding addiction (Bickel et al., 2007; Goldstein & Volkow, 2002a). For the individual struggling to overcome opioid dependence, the incentives to get high are mainly the relief from withdrawal and maybe feelings of euphoria, both of which are part of the immediate acute drug effects. By contrast, the benefits she might hope to attain by not using (repairing damaged relationships, improved health, etc.) would all take a long time to realize. The importance of this asymmetry in timing is self evident. If it were the other way around – if the reward or relief from using heroin didn't arrive for months, but all the benefits for abstaining arrived immediately, addiction would not be the problem it is (Monterosso & Ainslie, 2007; Monterosso et al., 2011).

A common perspective on addiction is closely related to the multi-agent competition conception of self-control struggle. In particular, addiction is viewed as not just linked to a high incentive value of the drug but also of a weakened general capacity to override impulses for immediate reward (Goldstein & Volkow, 2002a). While this might appear at first glance to be a theorized combination of strengthened System 1 and weakened System 2, there is little support or suggestion of a generalized increase in System 1 input. Indeed, incentive for non-drug rewards appears to be attenuated in association with addiction (Ahmed, 2005). But a generalized System 2 deficit is central to some views on addiction (Bechara et al., 2006).

Support for the idea of a generalized System 2 impairment in addiction can be found in the extensive literature on "executive function" capacities among individuals with addiction. Executive functions refer to a set of cognitive processes that help us control our thoughts and actions to achieve goals (e.g., working memory, inhibitory control, planning). Although the term can sometimes be used more inclusively, the capacities covered in the executive function literature are closely associated with System 2 processing, as both involve conscious, effortful control of our thoughts and actions and are heavily dependent on the functioning of the prefrontal cortex (especially lateral portions). Many studies suggest that "executive function" can be affected by chronic drug use. For instance, in studies with monkeys, long-term use of stimulants can lead to difficulties in adapting behavior to new situations (Jentsch & Taylor, 1999). Correlational studies offer some suggestion that long-term methamphetamine use in humans adversely impacts the capacity to inhibit motor responses (Monterosso et al., 2005; Salo et al., 2002). Moreover, there is considerable evidence that individuals struggling with addiction experience dopamine deficiency, with corresponding signaling deficits in areas of the prefrontal cortex that are important for decision-making and self-control (Koob & Volkow, 2010; Martinez et al., 2012; Volkow et al., 2009).

This evidence of impaired executive function in connection with addiction should not be overstated. Neuropsychological assessments of patients with substance use disorders generally do not show dramatic abnormalities (e.g., Dean et al., 2013), and there is no shortage of well-known public figures that maintain very high functioning, even as they struggle with addiction. That said, how might indices of executive functioning capacity inform answers to the question of whether an addictive behavior is free in the compatibilist sense? If you are trying to quit a heroin addiction, and if you experience a strong craving to use and then you relapse, "Did you do what you wanted to do?" The desire was there, but was it you? In her keynote address to the American Psychiatric Association, Nora Volkow called attention to evidence of impairment of the prefrontal cortex in association with addiction and argued that addiction entails "disruption of the areas of the circuits that enable us to exert free will, that enable us to exert free determination" (Volkow, 2015). The implicit picture here is that of a System 1 vs. System 2 competition, in which "you" is equated with System 2. Even after someone with an addiction wants to quit, System 1 will still generate urges to use, and so successful cessation requires a System 2 that can rein those urges in. If both systems are viewed as equally "you," then a shift in the balance of those systems would not have any implications on the extent to which the behavior was free in the compatibilist sense. But Volkow's perspective here, and we think it is a perspective shared by many (Baumeister et al., 2007; Bechara, 2005; Frankfurt, 1971; Goldstein & Volkow, 2002b), treats System 2 as more "you" than System 1. To the extent you agree with this point of view, neuroscience can establish whether free will is present, or perhaps the degree to which it is present, by investigating the output and capacity of the networks that instantiate this multi-agent system.

3.10 Final Remarks

Summarizing the previous arguments, we think that neuroscience cannot adjudicate free will in the libertarian sense, because the concept of *causa sui* has no place in science. But neuroscience could be used to assess whether behavior is free in the compatibilist sense. Indeed, theorists are already drawing such connections, suggesting addictive behavior is unfree because of its connection to habit control (Everitt & Robbins, 2005) or because of its basis in System 1 control (unchecked by System 2) (Volkow, 2015). Of course, it does not require understanding of the neural circuits relevant to habits, or to System 1 vs. System 2 valuation, to judge some behavior to be less free than other behavior. But the everyday bases for making such assessments are imprecise, and they can be made more difficult when those being judged have incentive to misrepresent their motivations. Views regarding which behaviors (and especially which socially undesirable behaviors) are free is an important source of disagreement in connection with culture wars (Everett et al., 2021). With progress in neuroscience, debates based on highly subjective grounds may give way to data-based evaluation of what behavior is and is not free.

References

Ahmed, S. H. (2005). Imbalance between drug and non-drug reward availability: A major risk factor for addiction. *European Journal of Pharmacology, 526*(1–3), 9–20.

Andersen, R., & Ross, M. (2019). The intention machine. *Scientific American, 320*(4), 24–31.

Anscombe, G. E. M. (1956). Intention. *Proceedings of the Aristotelian Society, 57*, 321–332. https://www.jstor.org/stable/4544583

Baumeister, R. F., Vohs, K. D., & Tice, D. M. (2007). The strength model of self-control. *Current Directions in Psychological Science, 16*(6), 351–355.

Bechara, A. (2005). Decision making, impulse control and loss of willpower to resist drugs: A neurocognitive perspective. *Nature Neuroscience, 8*(11), 1458.

Bechara, A., Noel, X., & Crone, E. A. (2006). Loss of willpower: Abnormal neural mechanisms of impulse control and decision making in addiction. *Handbook of Implicit Cognition and Addiction*, *1*, 215–232.

Belin, D., Jonkman, S., Dickinson, A., Robbins, T. W., & Everitt, B. J. (2009). Parallel and interactive learning processes within the basal ganglia: Relevance for the understanding of addiction. *Behavioural Brain Research*, *199*(1), 89–102.

Bickel, W. K., Miller, M. L., Yi, R., Kowal, B. P., Lindquist, D. M., & Pitcock, J. A. (2007). Behavioral and neuroeconomics of drug addiction: Competing neural systems and temporal discounting processes. *Drug and Alcohol Dependence*, *90*(Supplement 1), S85–S91. https://doi.org/10.1016/j.drugalcdep.2006.09.016

Bourget, D., Chalmers, D. J., & Chalmers, D. (2023). Philosophers on philosophy: The 2020 philpapers survey. *Philosophers' Imprint*, *23*. https://journals.publishing.umich.edu/phimp/article/id/2109/?utm_source=miragenews&utm_medium=miragenews&utm_campaign=news

Bredikhin, D., Germanova, K., Nikulin, V., & Klucharev, V. (2023). (Non)-experiencing the intention to move: On the comparisons between the Readiness Potential onset and Libet's W-time. *Neuropsychologia*, *185*, 108570.

Davidson, D. (2015). Freedom to act. In *Essays on freedom of action (Routledge revivals)* (pp. 137–156). Routledge. https://api.taylorfrancis.com/content/chapters/edit/download?identifierName=doi&identifierValue=10.4324/9781315719481-7&type=chapterpdf

Dawkins, R. (1996). *The blind watchmaker: Why the evidence of evolution reveals a universe without design*. WW Norton & Company. https://books.google.com/books?hl=en&lr=&id=sPpaZnZMDG0C&oi=fnd&pg=PR13&dq=the+blind+watchmaker&ots=tGl4aY8_5x&sig=ghAzyqkhuvqhinEkh2KRYO36qbU

Dean, A. C., Groman, S. M., Morales, A. M., & London, E. D. (2013). An evaluation of the evidence that methamphetamine abuse causes cognitive decline in humans. *Neuropsychopharmacology*, *38*(2), 259–274.

Dennett, D. C. (1984). *Elbow room: The varieties of free will worth wanting*. MIT Press.

Dickinson, A. (1985). Actions and habits: The development of behavioural autonomy. *Philosophical Transactions of the Royal Society of London. B, Biological Sciences*, *308*(1135), 67–78.

Dunbar, R. I. M. (1998). The social brain hypothesis. *Evolutionary Anthropology: Issues, News, and Reviews*, *6*(5), 178–190. https://doi.org/10.1002/(SICI)1520-6505(1998)6:5<178::AID-EVAN5>3.0.CO;2-8

Dunbar, R. I. M., & Shultz, S. (2007). Evolution in the social brain. *Science*, *317*(5843), 1344–1347. https://doi.org/10.1126/science.1145463

Ekstrom, L. W. (2005). *Libertarianism and Frankfurt-style cases*. https://academic.oup.com/edited-volume/28230/chapter/213280765?searchresult=1&utm_source=TrendMD&utm_medium=cpc&utm_campaign=Oxford_Academic_Books_TrendMD_1

Elton, A., Smith, C. T., Parrish, M. H., & Boettiger, C. A. (2017). Neural systems underlying individual differences in intertemporal decision-making. *Journal of Cognitive Neuroscience*, *29*(3), 467–479.

Everett, J. A. C., Clark, C. J., Meindl, P., Luguri, J. B., Earp, B. D., Graham, J., Ditto, P. H., & Shariff, A. F. (2021). Political differences in free will belief are associated with differences in moralization. *Journal of Personality and Social Psychology*, *120*(2), 461.

Everitt, B. J., & Robbins, T. W. (2005). Neural systems of reinforcement for drug addiction: From actions to habits to compulsion. *Nature Neuroscience*, *8*(11), Article 11.

Falcon, A. (2006). *Aristotle on causality*. https://plato.stanford.edu/ENTRIES/aristotle-causality/

Felsen, G., & Reiner, P. B. (2011). How the neuroscience of decision making informs our conception of autonomy. *AJOB Neuroscience*, *2*(3), 3–14.

Frankfurt, H. G. (1971). Freedom of the will and the concept of a person. *The Journal of Philosophy, 68*(1), 5–20.
Fried, I., Mukamel, R., & Kreiman, G. (2011). Internally generated preactivation of single neurons in human medial frontal cortex predicts volition. *Neuron, 69*(3), 548–562.
Gazzaniga, M. S. (1967). The split brain in man. *Scientific American, 217*(2), 24–29.
Gazzaniga, M. S. (2014). The split-brain: Rooting consciousness in biology. *Proceedings of the National Academy of Sciences, 111*(51), 18093–18094. https://doi.org/10.1073/pnas.1417892111
Gilden, L., Vaughan Jr, H. G., & Costa, L. D. (1966). Summated human EEG potentials with voluntary movement. *Electroencephalography and Clinical Neurophysiology, 20*(5), 433–438.
Goldstein, R. Z., & Volkow, N. D. (2002a). Drug addiction and its underlying neurobiological basis: Neuroimaging evidence for the involvement of the frontal cortex. *The American Journal of Psychiatry, 159*(10), 1642–1652. https://doi.org/10.1176/appi.ajp.159.10.1642
Goldstein, R. Z., & Volkow, N. D. (2002b). Drug addiction and its underlying neurobiological basis: Neuroimaging evidence for the involvement of the frontal cortex. *American Journal of Psychiatry, 159*(10), 1642.
Haidt, J. (2001). The emotional dog and its rational tail: A social intuitionist approach to moral judgment. *Psychological Review, 108*(4), 814.
Haidt, J. (2006). *The happiness hypothesis: Finding modern truth in ancient wisdom*. Basic Books.
Hare, T. A., Camerer, C. F., & Rangel, A. (2009). Self-control in decision-making involves modulation of the vmPFC valuation system. *Science, 324*(5927), 646–648. https://doi.org/10.1126/science.1168450
Heyman, G. M. (2009). *Addiction: A disorder of choice*. Harvard University Press.
James, W. (1887). . . . *The laws of habit*. D. Appleton. https://books.google.com/books?hl=en&lr=&id=Ba_kU8iqh0EC&oi=fnd&pg=PA431&dq=william+james+on+habits&ots=dL0OQnw1jP&sig=QMpqVvYCmbDf7G9ANgWZoB7uapA
Jentsch, J. D., & Taylor, J. R. (1999). Impulsivity resulting from frontostriatal dysfunction in drug abuse: Implications for the control of behavior by reward-related stimuli. *Psychopharmacology, 146*(4), 373–390.
Johansson, P., Hall, L., Sikström, S., & Olsson, A. (2005). Failure to detect mismatches between intention and outcome in a simple decision task. *Science, 310*(5745), 116–119. https://doi.org/10.1126/science.1111709
Johnson, M. K., & Raye, C. L. (1998). False memories and confabulation. *Trends in Cognitive Sciences, 2*(4), 137–145.
Kable, J. W., & Glimcher, P. W. (2007). The neural correlates of subjective value during intertemporal choice. *Nature Neuroscience, 10*(12), 1625–1633.
Kable, J. W., & Glimcher, P. W. (2010). An "as soon as possible" effect in human intertemporal decision making: Behavioral evidence and neural mechanisms. *Journal of Neurophysiology, 103*, 2513–2531.
Kahneman, D. (2011). *Thinking, fast and slow*. Farrar, Straus and Giroux.
Kane, R. (2018). Responsibility, luck and chance: Reflections on free will and indeterminism. In *Agency and responsiblity* (pp. 158–180). Routledge. https://www.taylorfrancis.com/chapters/edit/10.4324/9780429502439-11/responsibility-luck-chance-reflections-free-indeterminism-robert-kane
Koob, G. F., & Volkow, N. D. (2010). Neurocircuitry of addiction. *Neuropsychopharmacology, 35*(1), Article 1.
Lewis, M. (2017). Addiction and the brain: Development, not disease. *Neuroethics, 10*(1), 7–18.
Libet, B., Wright Jr, E. W., & Gleason, C. A. (1983). Preparation-or intention-to-act, in relation to pre-event potentials recorded at the vertex. *Electroencephalography and Clinical Neurophysiology, 56*(4), 367–372.

Luo, S., Ainslie, G., Giragosian, L., & Monterosso, J. R. (2011). Striatal hyposensitivity to delayed rewards among cigarette smokers. *Drug and Alcohol Dependence*, *116*(1), 18–23.

Martinez, D., Saccone, P. A., Liu, F., Slifstein, M., Orlowska, D., Grassetti, A., Cook, S., Broft, A., Van Heertum, R., & Comer, S. D. (2012). Deficits in dopamine D2 receptors and presynaptic dopamine in heroin dependence: Commonalities and differences with other types of addiction. *Biological Psychiatry*, *71*(3), 192–198.

McClure, S. M., Laibson, D. I., Loewenstein, G., & Cohen, J. D. (2004). Separate neural systems value immediate and delayed monetary rewards. *Science*, *306*(5695), 503–507. https://doi.org/10.1126/science.1100907

Metcalfe, J., & Mischel, W. (1999). A hot/cool-system analysis of delay of gratification: Dynamics of willpower. *Psychological Review*, *106*(1), 3.

Monterosso, J., & Ainslie, G. (2007). The behavioral economics of will in recovery from addiction. *Drug and Alcohol Dependence*, *90*(Supplement 1), S100–S111. https://doi.org/10.1016/j.drugalcdep.2006.09.004

Monterosso, J. R., Aron, A. R., Cordova, X., Xu, J., & London, E. D. (2005). Deficits in response inhibition associated with chronic methamphetamine abuse. *Drug & Alcohol Dependence*, *79*(2), 273–277.

Monterosso, J. R., Kalechstein, A., & Cordova, X. (2011). 13 If only the hangover preceded intoxication: An integration of behavioral economic and neuropsychological approaches to impulsive choice. *Neuropsychology and Substance Use: State-of-the-Art and Future Directions*, 407.

Monterosso, J. R., & Luo, S. (2010). An argument against dual valuation system competition: Cognitive capacities supporting future orientation mediate rather than compete with visceral motivations. *Journal of Neuroscience, Psychology, and Economics*, *3*(1), 1–14.

Nahmias, E. (2011). Why 'willusionism' leads to 'bad results': Comments on Baumeister, Crescioni, and Alquist. *Neuroethics*, *4*(1), 17–24. https://doi.org/10.1007/s12152-009-9047-7

Nahmias, E. (2014). *Is free will an illusion? Confronting challenges from the modern mind sciences*. https://direct.mit.edu/books/edited-volume/chapter-pdf/2272144/9780262321488_caa.pdf

Nisbett, R. E., & Wilson, T. D. (1977). Telling more than we can know: Verbal reports on mental processes. *Psychological Review*, *84*(3), 231.

Palmer, D. (2014). *Libertarian free will: Contemporary debates*. Oxford University Press. https://books.google.com/books?hl=en&lr=&id=vKM-BAAAQBAJ&oi=fnd&pg=PP1&dq=libertarian+freewill+causa+sui&ots=NkhnCHoeb8&sig=ZfGAbzRDDn8g4SXkwmeiSgS-Ckk

Robbins, T. W., & Costa, R. M. (2017). Habits. *Current Biology*, *27*(22), R1200–R1206.

Salo, R., Nordahl, T. E., Possin, K., Leamon, M., Gibson, D. R., Galloway, G. P., Flynn, N. M., Henik, A., Pfefferbaum, A., & Sullivan, E. V. (2002). Preliminary evidence of reduced cognitive inhibition in methamphetamine-dependent individuals. *Psychiatry Research*, *111*(1), 65–74.

Sanford, P., Lawson, A. L., King, A. N., & Major, M. (2020). Libet's intention reports are invalid: A replication of Dominik et al. (2017). *Consciousness and Cognition*, *77*, 102836.

Santas, G. (Ed.). (2006). *The Blackwell guide to Plato's republic* (1st ed.). Wiley. https://doi.org/10.1002/9780470776414

Shefrin, H. M., & Thaler, R. H. (1992). *Mental accounting, saving, and self-control*. https://psycnet.apa.org/record/1993-97149-012

Soon, C. S., Brass, M., Heinze, H.-J., & Haynes, J.-D. (2008). Unconscious determinants of free decisions in the human brain. *Nature Neuroscience*, *11*(5), 543–546.

Strawson, P. F. (1974). *Freedom and resentment, and other essays*. Egmont Books.

Street, F. (2017, February 13). *Who's in charge of our minds? The interpreter*. Farnam Street. https://fs.blog/michael-gazzaniga-the-interpreter/

Trobisch, D. (2000). *The first edition of the new testament*. Oxford University Press. https://books.google.com/books?hl=en&lr=&id=kLDyPGJldl8C&oi=fnd&pg=PP2&dq=new+testament&ots=1LIFTXqPy_&sig=54XIoFt2eE5TKmb1nf8_2je59v0

Volkow, N. D. (2015). *Addiction is a disease of free will*. https://archives.drugabuse.gov/about-nida/noras-blog/2015/06/addiction-disease-free-will

Volkow, N. D., Fowler, J. S., Wang, G. J., Baler, R., & Telang, F. (2009). Imaging dopamine's role in drug abuse and addiction. *Neuropharmacology*, *56*(Suppl. 1), 3–8.

4

Neurolaw: Overview, Past, Present, Future

Emily R.D. Murphy

4.1 Introduction

Neurolaw started for me in the spring of 2003. Renowned psychologist Stephen Kosslyn and criminal and constitutional law professor Alan Dershowitz offered a small seminar class called "The Neuroscience of Law: Can a Legal System Be Grounded in Knowledge About the Brain?" The course catalog tantalized: "Discusses implications of neuroscience on the characterization of human nature, the relation between genes, brain, and social interactions, and determination of right vs. wrong." Surely we could cover such topics in a single semester, meeting once a week. Ten graduate or upper-level undergraduate psychology students and ten law students enrolled in the class and were paired up to self-design readings for each week.

The first class began with a clip from the 2002 movie *Minority Report*, with the screenplay loosely based on Phillip K. Dick's 1956 novella of the same name. The movie, in which people are arrested for crimes not yet committed but predicted by "precog" beings, posed a question that became the premise of the class: if we could use the brain to perfectly predict behavior, what result for the law? The class was a lot of fun but logistically a mess. There was not yet really a body of literature for students to select coherent readings from. There was a lot of riffing on the nature of human nature. I don't think it was ever repeated.

I happened to take the class, then go on to enroll in graduate school in neuroscience, at the same time the interdisciplinary field of "law and neuroscience" or "neurolaw" was really getting started. Some would trace the origins of "neurolaw" back to the 1990s, when advances in neuroimaging allowed researchers to map brain functions and analyze brain injuries in new ways. With brain imaging came the possibility of new evidence making visible and concrete what would previously have been invisible to a jury: the contents of someone's head. But what we presently understand as the field of "neurolaw" goes well beyond evidence of injury in tort law. As evidenced by the class I took, modern neurolaw started out by heading straight for old, fundamental questions about human nature, mental states, and responsibility.

 DOI: 10.1201/9781032717791-4

This chapter briefly traces the past, present, and future of "neurolaw." It is a testament to the growth of the field that space precludes a comprehensive survey. Thus, the history is predominantly a recounting of the field as I have experienced it, with attention given to particularly salient moments and publications that necessarily are influenced by that perspective. Predictions are heavily influenced by my own research agenda.

Here is the arc of the field as I see it: early neurolaw work operated on the premise that being able to look inside the brain would provide answers to some of the oldest, hardest questions that law faced. Thinkers and scholars focused predominantly on topics like the neuroscience of criminal responsibility and prediction of antisocial behavior – picking up from a long history of bio-criminology and building on excitement in behavioral genetics. Closely related to questions of responsibility were those of mind reading and especially brain-based lie detection: can we (finally) see what's going on in someone's mind by looking directly into their brain?

Over the past two decades, neither problem has been solved. Instead, neurolaw scholarship expanded to address a wider range of more pragmatic legal questions from juvenile justice to sentencing, with most work still focused on courtroom adjudication and prevention of societal harm by addressing antisocial behavior at the level of the individual.

Current interesting neurolaw debates continue to grapple with the legal and ethical implications of new neuroscientific and behavioral understanding, accumulation of brain and behavioral data, and technological advancement. Key issues include the capabilities and limits of brain imaging technologies to predict behavior, and neuroethics of enhancement, neurodiversity, and privacy. The introduction of neuroscientific evidence in courtrooms continues to raise concerns about misinterpretation and over-reliance on brain scans, as well as the problem of making inferences about individuals from group-based data. Less interesting, but still prevalent: the free will and criminal responsibility debate persists – and probably always will – with repackaging of old arguments in new technology.

What has been accomplished? Certainly neuroscience is showing up more and more in courtrooms, with mixed impacts. Judicial education about brain science and its limitations continues to expand every year. But what of the theoretical marriage between brain function and fundamental legal concepts? Some might say that finding neural correlates of complex behavior has led to a lot of: so what? What do we learn from brain correlates that is both legally meaningful and not already revealed by behavior?

Focusing on the neuroscience of individual behavior and on the law as expressed in the courtroom has not, and almost certainly will not, entirely reshape the part of the legal system that adjudicates and assigns culpability to that behavior. Repeated discussions about free will, responsibility, and nature versus nurture have not, and probably will not, lead to a revolution in criminal law. So, if neurolaw is really asking "can a legal system be grounded in knowledge about the brain?", it benefits from taking a broader view of what the legal system is, beyond the courtroom and beyond the prediction or adjudication of individual behavior.

Recent neurolaw work has started to reflect this shift, taking law as how we collectively govern ourselves and allocate societal benefits, protections, and responsibilities. Law is a social determinant of health and wellbeing, not only a mechanism of dispute resolution or adjudication of blameworthiness for past conduct. Beyond using neuroscience as a tool for the law, especially as a form of evidence, present debates consider whether law and policy should adapt to reflect understandings of how our brains work, how we can monitor or modify them, and how they thrive. This work is also starting to shift in response to the way neuroscience research is advancing. We know enough now to know that brain and behavior relationships are complex, subtle, and extremely "noisy" at the level of the individual. Large data sets are necessary for research to have sufficient power to find small effects.

Looking ahead, the future of neurolaw is turning towards structural and systemic issues concerning human well-being and the production of diverse, self-governing societies – the promotion of the means of thriving in the current age of technology, inequality, and systemic threats. The future of neurolaw should push beyond the current individualistic boundaries of "behavioral public policy" – getting individuals to behave a certain way within an existing system – and press forward on the systems themselves. The future of neurolaw should use legal rights frameworks to protect and promote cognitive liberty (Farahany, 2023), claim and protect a right to brain health, and build tools to assess public policies and legal rules based on their impact on "collective cognitive capital" as a valuable resource (Murphy, 2022). Law and policy should aim to enhance collective brain health and rights to self-determination because both are essential to promoting the values and practices of autonomy, self-governance, and widespread human flourishing. These proposed future applications of neurolaw move the field in a more structural, society-focused direction. They also help inform brain and behavioral science research agendas.

In summary, this chapter introduces key issues, debates, and questions in the still-young, dynamic field of neurolaw. It traces major developments in neurolaw from its origins to current controversies to future possibilities at the intersection of neuroscience and law.

4.2 Neurolaw's Start as Criminology and Philosophies of Criminal Responsibility

Advances in understanding our biology have always raised questions about human nature. For neuroscience, the possibility of understanding the brain seems to directly address difficult problems that the law confronts resolving: what is (or was) going on in someone's mind? New brain imaging techniques offered the option to look inside at the brain and try to answer this question.

Prior to the 1990s, studying the brain was largely limited to post-mortem anatomical investigation or animal studies, with limited applications to humans. The development of positron emission tomography (PET) scanning in the late 1980s, and then functional magnetic resonance imaging (fMRI) in the early 1990s, opened up profound new possibilities for understanding the living, working human brain.

PET and fMRI allowed researchers to observe the living brain in action. For the first time, we could map out which brain regions were active when a person engaged in particular mental tasks – from reading or doing math problems to wrestling with a moral dilemma, feeling afraid, or even meditating or praying. Structural MRI also enabled much more granular analysis of brain abnormalities or injuries.

This ushered in a new era for neuroscience but also immediately posed questions for the legal system. If we could now literally see what the brain was doing during decision-making, did this change the notion of personal responsibility embedded in the law? Did neuroimaging reveal that some violent offenders had brain deficits that prevented them from controlling their behavior – calling into question their criminal culpability? Could brain scans detect if someone was lying? That is, neuroscience seemed to offer new possible solutions to very old questions that law grappled with.

Early subjects for this approach of applying new technology to old questions were persons accused of violent crimes. Continuing in a long tradition of biological criminology, Adrian Raine and colleagues started using PET to assess the brains of people accused of murder and other violent crimes in the mid-1990s (1994, 1997). No one yet called this "neurolaw," but it laid the foundation for what would become early neurolaw's fixation on crime and antisocial behavior. (Though I am critical of this ongoing focus of neurolaw, this was also my own entry: violence is what turned me towards the study of psychology and human behavior in the first place. My assigned first year undergraduate writing course was "Violence and Nonviolence in Theory and Fiction." The course redirected my undergraduate major from biochemistry to psychology and mind/brain/behavior. At some point after that class, I started reading the articles about the brain PET scans of convicted murderers done by Adrian Raine, reporting reduced glucose metabolism in their prefrontal cortices.)

The 1990s and early 2000s were a time of excitement and anxiety about behavioral genetics and neuroscience. In the early 1990s, the Human Genome Project also launched the field of study about ethical, legal, and social implications (ELSI) of genetics (Dolan et al., 2022). The 1990s was also announced as the "Decade of the Brain" (Goldstein, 1990), though the formal brain science analog to genetics' ELSI was not termed "neuroethics" until a germinal conference hosted by the Dana Foundation in May 2002 (Marcus and Dana Foundation, 2002). Both fields, and their potentially science-fiction-esque implications, captured the public imagination. In 2002, *The Economist* teased in an article titled "Open Your Mind" that "Genetics may yet threaten privacy, kill autonomy, make society homogeneous and gut the concept of human nature. But neuroscience could do all of these things first." In the early 2000s, the genesis of modern neurolaw was getting underway.

That *Economist* article was the lede in the 2004 report of an invitational meeting titled "Neuroscience and the Law: Brain, Mind, and the Scales of Justice" convened by the

American Association for the Advancement of Science and the Dana Foundation. This seems to have been the first major conference specifically about the legal implications of neuroscience. The first topics considered by the AAAS/Dana report were "competency, free will, and the genesis of violent behavior" (Garland, 2004). Not all of the discussion at that meeting focused on criminal behavior and responsibility; Hank Greely's essay focused on litigation more generally, as well as mental privacy concerns (2004). But the dominant themes in the early years of neurolaw were focused on the causes of and decisions about individual human behavior, almost exclusively in the criminal context.

In late 2004, the London-based Royal Society published a special issue titled "Law and the Brain." The introductory essay argued that the purposes of law "depend on collective judgments about human motivation, the factors that influence human beings to think, decide and act as they do" and that accurately resolving problems involving questions of human motivation will be assisted by mechanistic understanding of the brain (Bingham, 2004). Of the sixteen articles published in that issue, two had particularly large impact. The first, by neuroscientist/philosopher Josh Greene and neuroscientist Jonathan Cohen, was titled "For the Law, Neuroscience Changes Nothing and Everything." In the 20 years since its publication, the article has been cited over 1100 times.

Greene and Cohen argued, in essence, that brain science would change the (criminal) law by shifting the general public's moral intuitions about free will and responsibility. Consequently, this change in perspective would lead to a shift in punishment away from retribution towards consequentialism (Greene and Cohen, 2004). Greene and Cohen's prediction was and is fundamentally an empirical question about how brain science would change cultural mindsets. A 2013 review of the evidence of whether the general public's understanding of free will and personhood has been affected by neuroscientific ideas found that this prediction was not coming true (O'Connor and Joffe, 2013). Instead,

> the bulk of the evidence . . . suggests that claims that neuroscience will dramatically alter people's relations with their selves, others and the world are overstated. In many cases, neuroscientific ideas have assimilated in ways that perpetuate rather than challenge existing modes of understanding.
>
> **(p. 262)**

In other words, neuroscientific information does not appear to have yet changed peoples' minds about their personhood or sense of free will and responsibility.

Back to 2004: in the same issue, neuroscientist Robert Sapolsky (who now self-identifies as the "lunatic fringe in terms of the views where the two fields intersect") (2021) argued for the "hard incompatibilist" position for neurolaw. The crux of his position is that our brains are mechanistic and therefore there can be no such thing as "fault" because "what we call free will is the biology that has not been discovered yet" (Sapolsky, 2021, 2004). His article, titled "The Frontal Cortex and the Criminal Justice System," has been cited over 240 times. And Sapolsky's influence as a public intellectual undoubtedly helped get the attention of major philanthropic funders, who substantially invested in neurolaw starting in 2007, as discussed below.

Notably, neither of these two highly influential articles at the dawn of "neurolaw" were written by legally trained scholars. The preeminent legal scholar addressing issues of free will and responsibility was Stephen Morse, who took a decidedly different perspective. Morse argued in "New Neuroscience, Old Problems" in the 2004 AAAS/Dana conference proceedings that unless new neuroscience fundamentally shifted our self-concept, the law as is could incorporate brain science's findings rather than demanding radical revision. He had made many of these points as far back as 1996, in an article called "Brain and Blame" (Morse, 1996). Morse's body of work in law, psychology, and neuroscience consistently explains that different kinds of explanations can be given for behavior, such as mechanistic explanations and reasons-giving explanations, and that generally the law's concept of the person is "a practical reasoning, rule-following being, most of whose legally relevant movements must be understood in terms of beliefs, desires, and intentions" (Morse, 1996). That is, contrary to Sapolsky's view, the law does not essentially require free will; it requires that behavior be "reasons responsive" in order to be culpable. Mere causation is an explanation for behavior but not per se an excuse for that behavior, because excusing conditions are behavioral and normative rather than materialist/mechanistic. Where the law is concerned with mechanistic, causal accounts of behavior (such as in affirmative defenses of insanity), it is at the margins – where responsiveness to reasons has broken down entirely, and the evidence of the broken brain mechanism supports the behavioral/normative conclusion of a lack of legal capacity for responsibility. Morse does, in essence, agree with Greene and Cohen that what fundamentally drives legal conceptions of responsibility are normative conceptions of ourselves, and if those are truly altered by neuroscience, then the legal system may require deeper rethinking. But he also diagnosed "Brain Overclaim Syndrome" in an influential 2006 article, which he identifies as "mak[ing] claims about the relation of the brain to responsibility that cannot be sustained logically or empirically," largely due to a misunderstanding of the legal criteria for responsibility: "To believe that brain evidence has more than simple evidentiary value for assessing responsibility is to misconceive the criteria for responsibility" (Morse, 2006).

But the cases at the margins continued to draw public attention to the burgeoning field of neurolaw. In March 2007, the *New York Times Magazine* published a long article by legal commentator Jeffrey Rosen that began "When historians of the future try to identify the moment that neuroscience began to transform the American legal system, they may point to a little-noticed case from the early 1990s." The case was Herbert Weinstein's prosecution for his wife's murder by defenestration and his brain-based defense that he should not be deemed responsible due to an arachnoid cyst in his brain. This case, sometimes given the pseudonym of "Spyder Cystkopf," has become one of a handful of case studies that, in early law and neuroscience, were often cited as raising profound or considerable implications for law more generally. Another was the case of the schoolteacher-turned-sex offender whose deviant behavior subsided with the removal of a large brain tumor on his orbitofrontal lobe – and returned when it regrew (Seiden,

2004). Sensational cases can sometimes lead to profound changes in the legal system – the 1603 treason trial of Sir Walter Raleigh is widely cited as the motivating origin for, variously, the rule against hearsay and the right to confront one's accusers. But, in line with Stephen Morse's prediction, neuroscience has not yet similarly "transformed" the American (criminal) legal system.

Where neuroscience has arguably had the most significant impact on the substance of the legal system is in the area of juvenile justice. In a series of cases in the 2000s, the U.S. Supreme Court drew on neuroscience research showing that the adolescent brain is still maturing to place constitutional limits on harsh punishments for juveniles. In *Roper v. Simmons* (2005), the Court outlawed the death penalty for crime committed by minors. In *Graham v. Florida* (2010), the Court ruled that sentencing juveniles to life without parole for non-homicide crimes violated the 8th Amendment. In *Miller v. Alabama* (2012), the Court held that mandatory life-without-parole sentences were unconstitutional for all crimes committed by minors. The Court reasoned, in part, that youths' still developing brains made them less culpable and more amenable to reform. Many observers have argued that the neuroscientific evidence, mostly presented in "friend of the court" amicus briefs and alluded to in parts of the opinions, persuasively influenced some justices not because it revealed new information but because it supported and aligned with common sense about adolescent development (Steinberg, 2013, 2017). But others have argued that the brain science distracted from appropriate legal reasoning about the relationship between capacity and culpability (Brooks, 2014).

However, the Supreme Court's invocation of brain science also highlighted a key tension defining neurolaw debates – the potential for neuroscience to be misinterpreted, overapplied, or reductively applied in legal contexts. For example, Terry Maroney argued that neuroscience in juvenile court cases offered a "false promise" and that incorporating the science into juvenile justice law would be unlikely to help individual defendants, and was more properly considered by legislatures rather than courts (2013). This is primarily because neuroscience simply showed that on average adolescents' brains are less mature, not that any *individual* youth's brain was less capable, rendering them less culpable. Moreover, where scientists perceived a reductiveness and loss of nuance in Supreme Court pronouncements about what the science said about capacities of adolescents, skeptically-minded lawyers (exemplified by Justice Antonin Scalia) perceived inconsistency. Following on Justice Kennedy's similar questioning at oral argument, Justice Scalia's dissent from the *Roper* majority critiqued the American Psychological Association's position in the *Roper* case as inconsistent with its prior position in a case regarding adolescent access to abortion with respect to juvenile moral responsibility:

> [T]he American Psychological Association (APA), which claims in this case that scientific evidence shows persons under 18 lack the ability to take moral responsibility for their decisions, has previously taken precisely the opposite position before this very Court. In its brief in *Hodgson v. Minnesota*, 497 U. S. 417 (1990), the APA found a "rich body of research" showing that juveniles are mature enough to decide

> whether to obtain an abortion without parental involvement. . . . The APA brief, citing psychology treatises and studies too numerous to list here, asserted: "[B]y middle adolescence (age 14–15) young people develop abilities similar to adults in reasoning about moral dilemmas, understanding social rules and laws, [and] reasoning about interpersonal relationships and interpersonal problems."
>
> **(*Roper v. Simmons*, Justice Scalia, dissenting, pp. 11–12)**
> **cited in (Steinberg et al., 2009)**

Of course, the epistemic value of the brain science data for either purpose depends enormously on the "fit" between the behavior being assessed and, relatedly, the scientific literacy of judges gatekeeping it. Conflating two entirely different types of behavior – whether to give birth versus whether to hurt someone in the heat of the moment surrounded by peers – into a single measure of "maturity" demonstrates a prime example of the challenges of integrating nuanced scientific findings into coherent, consistent legal theories about responsibility and capacity of a particular subgroup of people. (Maslowsky et al., 2024)

Overall, 2007 was a major year for neurolaw. In October 2007, the John D. and Catherine T. MacArthur Foundation launched the interdisciplinary Law and Neuroscience Project to bring together neuroscientists, law professors, philosophers, and practitioners in each field. It was Robert Sapolsky who proposed to the MacArthur Foundation that the foundation "go big" on law and neuroscience (Jones, 2022). The initial $10 million grant was the first systematic effort to integrate law and neuroscience and was focused on the realm of criminal justice. The initial project had three working groups on the topics of addiction, brain abnormalities, and "normal" decision making (as it related to concepts such as criminal responsibility). (I was one of the first set of postdocs on the project.)

The first phase of the Law and Neuroscience Project "pursued several over-arching questions at the intersection of neuroscience and law" focused on courtroom adjudication, criminal responsibility, and prediction of antisocial behavior. The Law and Neuroscience Project coincided with an acceleration in the pace of neurolaw related publications, as shown in Figure 4.1. The project accomplished a number of its stated objectives, perhaps most tangibly in the area of brain science as courtroom evidence.

4.3 Neurolaw's Past, Present, and Future as Courtroom Evidence

At its modern start, the biggest controversy in law and neuroscience was essentially the battle between those (mostly neuroscientists) who thought that a materialistic understanding of behavior would undermine free will and thus undercut the foundations of the criminal legal system and those (mostly lawyers) who thought that this concern was greatly exaggerated and that the legal system would persist in using its

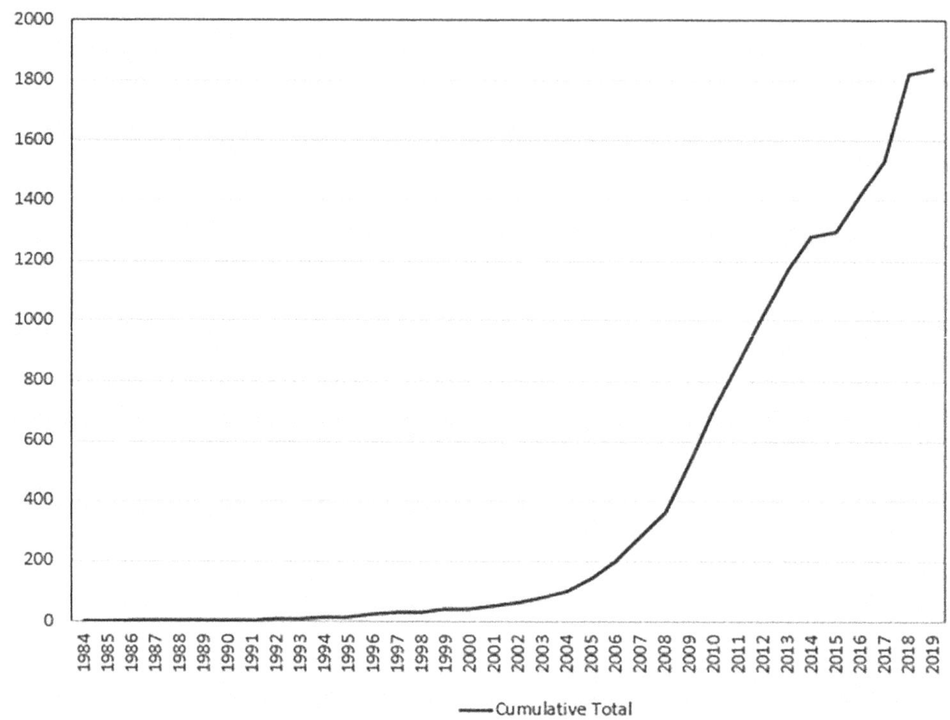

Figure 4.1 Cumulative total of law and neuroscience publications from 1984–2019, as compiled by the MacArthur Foundation Research Network on Law and Neuroscience, www.lawneuro.org/bibliography.php.

phenomenological understanding of human behavior. Shared across these camps were concerns about brain science showing up as *evidence* of subjective internal states. Adjacent to moral and philosophical concerns about legal responsibility, these fact-finding questions addressed hard and persistent problems in the nuts and bolts of adjudication, and the empirical and normative work in these areas continues to become more sophisticated.

The advent of neuroimaging created a new form of evidence, making visible the previously invisible workings inside someone's head. A range of courtroom applications followed in both criminal and civil cases, as well as attempts to develop applications for forensic or courtroom use such as lie detection and memory detection. One of the most difficult tasks that the law assigns to a fact-finder (be it a jury or a judge) is essentially that of mind-reading: what was a person's intent, awareness, or other legally relevant mental state at the time of the action being adjudicated? That is: the law already requires fact-finders to engage in time travel and mind-reading all the time (Brown and Murphy, 2010; Murphy and Greely, 2011). Could neuroscience help them in their task?

There is now a rich literature examining the uses as well as risks of neuroscience evidence in the courtroom. Early observers included critics as well as promoters; a student note law review article from 1997 presciently criticized the use of "hard" neuroimages to support "soft science" psychiatric diagnoses in court (Kulynych, 1997). Anthropologist Joe Dumit demonstrated how PET scans were both constructed by human choices and in turn influencing conceptions of personhood in social applications such as law, leading to his observation of our increasing dependence on science as a source of social authority (2004). One emergent theme of concern was that neuroscience evidence in capital sentencing would prove to be a "double-edged sword": even if offered to mitigate responsibility, it would also indicate a potential for future dangerousness (Barth, 2007; Snead, 2008; Mackintosh, 2011; Aspinwall et al., 2012; Jones and Shen 2012). But Deborah Denno's recent study of ~800 criminal cases from 1992–2012 suggests that such concern is overblown in that most uses are for mitigation purposes and that in fact defense counsel's failure to pursue such evidence may be the basis for an ineffective assistance of counsel claim (2015).

On the ground, neuroscience is entering the courts more and more each year. Nita Farahany conducted a peer-reviewed empirical analysis of the use of neurobiological evidence in criminal cases in the use from 2005 to 2012, finding dramatic growth in both the number of cases and the substantiveness of the discussion of the evidence (2015). Contrary to dominant assumptions among neurolaw scholars at the time, she found that most (60%) of cases in which neuroscience evidence was introduced were not capital cases. About 44% of the neurobiological claims raised were attempts to mitigate sentencing; nearly half of those were arguing that a defendant received constitutionally ineffective assistance of counsel by failing to introduce brain evidence at sentencing. A small proportion of cases in the study pertained to juveniles, where advocates used a more general "developing brain" theory rather than making individualized claims like those seen for adult offenders. Nevertheless, Farahany concluded that increasing presence of neurobiological evidence in courts "continues to be haphazard, ad hoc, and often ill conceived."

As evidence, brain science can be relevant to a legal issue in a number of ways: buttressing and explaining behavioral evidence, detecting legally relevant facts, predicting future behavior or response to interventions, and providing interventions (Jones and Wagner, 2019). The utility of neuroscience evidence in any of these categories is often a mixed question of law and fact, depending on the strength of mechanistic, logical, and normative inferences. Admissibility is contextual, too; legal scholars have outlined the parameters of admissibility of such evidence, which has often been the stated goal of many researchers of forensic uses of the technology (Moriarty, 2008; Murphy and Rissman, 2020; Moriarty, 2021). The next few sections look at a few specific evidentiary uses of brain science as evidence, each the subject of a rich literature.

Lie detection, "brain fingerprinting," and memory detection. The U.S. legal system has a long history with technological lie detection. A 1923 murder case involving evidence of a "systolic blood pressure deception test" served for decades as the legal standard for assessing expert testimony (*Frye v. United States*, 1923). A 1998 Supreme Court case involving a defendant's attempt to admit polygraph evidence in military court reiterated that 1) there was no scientific consensus on the reliability of polygraphs, and 2) the question of a witness's credibility was ultimately a decision for the jury (*United States v. Scheffer*, 1998). But the advent of brain imaging nearly immediately raised the specter of brain-based lie detection, on the theory that where physiological measures of deception were unreliable or could be defeated with countermeasures, getting right to the source of the lie would be more accurate and harder to beat.

Early brain-based work used electroencephalogram (EEG) technology focused on the P300 signal and variations of a "guilty knowledge test" termed "brain fingerprinting" (Farwell and Donchin, 1991; for a review, see Murphy and Rissman, 2020). The major proponent of this work and its commercial applications has claimed that it is both admissible in a U.S. court and was instrumental in overturning a conviction. These claims are overstated as a matter of legal accuracy; the conviction was ultimately overturned by a state supreme court for reasons unrelated to the brain fingerprinting test (Greely and Illes, 2007). But EEG-based memory detection made international headlines in 2008 when a woman in India was convicted of murder of her fiancé based in part on evidence that her brain activity demonstrated "experiential knowledge" of the crime (Giridharadas, 2008; for a review of the technology and the legal situation in India, see Murphy and Rissman, 2020).

Can brain information detect whether someone has such autobiographical memory? In 2007, Michael Gazzaniga predicted that in ten years neuroscientists could possibly show brain differences when people testify about their own prior acts versus acts they witnessed (Rosen 2007). Indeed, this seems to be somewhat true – but one major problem is that false memories that are incorrectly but subjectively perceived as true have a similar brain activation pattern (as assessed by functional magnetic resonance imaging) to true, autobiographical memories. (See Murphy and Rissman, 2020). Notwithstanding the efforts of researchers driven by agendas of legal admissibility, EEG-based memory detection has yet to become accepted as forensic evidence in U.S. courts.

Memory detection overlaps with deception (lie) detection, which also became a focus of brain imaging research. Early fMRI studies tantalized at the prospect of finding lies in the brain (Spence et al., 2001; Lee et al., 2002; Langleben et al., 2002, 2005; Kozel et al., 2004, 2005). Two companies started offering their services for fMRI-based lie detection or "truth verification" (Greely and Illes, 2007). No Lie MRI started offering services in August 2006, proposing potential applications in dating, relationships, and legal liability. Its first reported customer "sought to use their technology to rebut claims of arson and insurance fraud" (Greely and Illes, 2007). CEPHOS Corporation also offered lie detection services starting around 2007, claiming greater than 90% accuracy for its technology. The services and products of CEPHOS became the subject of evidentiary hearings in the 2010 federal criminal trial *United States v. Semrau*, where they were offered by the defendant accused of Medicare/Medicaid fraud as evidence that he had not knowingly engaged in prohibited billing practices (Jones and Shen, 2012). After a pre-trial evidentiary hearing, the magistrate judge found (and the trial judge later adopted) that "at least at this early stage in its development, fMRI-based lie detection does not satisfy the requirements of [Federal Rule of Evidence] 702," the legal rule governing the admissibility of expert testimony (*United States v. Semrau*, 2010). The case garnered significant national media attention, and the trial court's decision was later upheld by the Court of Appeals for the Sixth Circuit (*United States v Semrau*, 2012). A New York State court similarly rejected fMRI lie detection evidence from CEPHOS in a civil case when offered by a plaintiff to prove a key witness's credibility in a case about unlawful retaliation in an employment context (*Wilson v. Corestaff Servs.*, 2010). The court excluded the evidence on the grounds that credibility was a determination for the jury to make (and thus that the technology would not help them) but also noted that under the relevant standard for assessing reliability of scientific evidence, "even a cursory review of the scientific literature demonstrates that the plaintiff is unable to establish that the use of the fMRI test to determine truthfulness or deceit is accepted as reliable in the relevant scientific community."

As of the writing of this chapter, CEPHOS Corp seems to have pivoted away from forensic fMRI towards forensic DNA testing. No Lie MRI no longer seems to be a going concern. While one of the MacArthur Law and Neuroscience Project's initial aims was to "Enable more reliable identification of deceptive testimony," excitement and discussion about brain-based lie detection in court has died down significantly since the *Semrau* case.

Pain and privacy. Perhaps more ubiquitous in the legal system than concerns about lying is the presence (or absence) of pain, and its severity. Law professor Amanda Pustilnik has asked "What would the law do differently if it could see pain, as is increasingly possible through new neuroimaging technologies?" (2015). Pustilnik outlined how brain-based evidence of pain should be admissible in court to show general biology and effects of chronic pain but is not yet sufficiently precise to prove individual pain. She also argued that it should change the law in areas such as social security disability determinations, as well as how the fact of pain neuroimaging reveals that pain generally plays a role in law broadly as a heuristic for values and thus an avenue to impact legal doctrine more broadly (Pustilnik, 2012). General consensus in the field is still that, while research continues, the use of brain imaging as legal evidence of a given individual's pain is not yet ready for the courtroom (Davis et al., 2017). But if such evidence becomes sufficiently accurate and reliable, some empirical evidence indicates that it might be persuasive to fact-finders in proving pain and its severity and assessing how it should be compensated (Phalen et al., 2021).

The possibility of brain-based pain detection has raised deeper questions about mind-reading in law more generally (Greely, 2015). Adam Kolber predicted in 2007 that such efforts, properly understood as making subjective experiences into objective evidence, raised manageable privacy concerns because while pain detection would look into someone's subjective experience, that intrusion would be limited with respect to other thoughts. Amanda highlighted that pain detection in adjudication also sweeps in a vulnerable population, adding further risks of subjugation via monitoring and measurement to a group already burdened by lower social status (2018). And while pain-detection techniques are mostly contemplated for voluntary use by plaintiffs seeking additional proof of injury or disability, the prospect of government-controlled neuroimaging has raised challenging questions about the scope of individual constitutional rights against unlawful searches and seizures and the potential for thoughts to be self-incriminating (Farahany, 2012a, 2012b).

Predictions and interventions. Not only does the law need to read minds all the time, but it also is asked to assess risk, predict the future, and fix or prevent problems. Hearkening back to *Minority Report*, one of the MacArthur Project's first set of aims was to "Aid law's abilities to predict recidivism and future dangerousness." Some of the work supported by the Project suggested that brain activity during a certain impulse-control behavioral task predicted rearrest among adult offenders (Aharoni et al., 2013). The findings generated significant popular press and scholarly pushback. Further analysis of that data led the authors to re-affirm the findings and argue for the utility of neurobiological assessments in predicting rearrest (Aharoni et al., 2014). A different set of researchers, some of whom were also affiliated with the MacArthur Project, responded with a critical assessment of the state of neuroprediction, arguing that best practices would include pre-registration of study designs, hypotheses, and outcome measures; validation using independent samples; and the development of appropriate norms using large sample sizes across a range of demographics (Poldrack et al., 2018). Actuarial prediction in legal contexts is the subject of an enormous (and critical) literature, and it remains to be seen whether the practical, pragmatic, and ethical difficulties of adding brain-based information to any prediction mechanisms would be effective enough to be worth it.

Advances in brain imaging have also invited advances in brain intervention as form of either punishment or therapeutic justice, though at least in the United States such uses remain in the realm of science fiction. These are not uses of brain science as evidence, but they do fit in the context of being administered by the courts. In a 2007 *New York Times* article, Jeffrey Rosen quoted Robert Sapolsky as speculating how "damaged" brains could be repaired as a form of therapeutic justice. Neuroethicist Paul Root Wolpe identified the possibility of using transcranial magnetic stimulation to "turn off or inhibit" a discrete part of the brain "operative in violent or sexually predatory behavior" (Rosen, 2007). Such speculation seems to have died down, probably because no such "discrete" part of the brain has materialized. Other therapeutic justice possibilities arise from psychopharmaceutical interventions, particularly in areas of treatment of addiction and co-morbid behavioral disorders. Potential introduction through an overburdened system of "problem solving courts," adjacent to the regular criminal legal system, presents both risks to individual autonomy as well as institutional competency and legitimacy (Murphy, 2013). Finally, some scholars have begun to grapple with the legal and ethical implications of implantable brain devices as a form of rehabilitation (Gilbert and Dodds, 2020; Ligthart et al., 2021). These debates are taking place in the present, but any such actual application seems firmly a challenge for the future.

Use and impact of neuroscience as evidence: G2i and persuasion. The drive to use brain science as evidence raises two meta-questions shared by other forms of scientific evidence as applied to legal settings. First is the problem of "group to individual" inference (or "G2i"): whether one can make definitive determinations about a particular individual/circumstance from generalized, group-based data (Faigman et al., 2014). The second is whether brain science has exceptionally persuasive power over judges and juries beyond what is warranted by its evidentiary strengths.

Work on the "G2i problem" is an area of real conceptual strength arising out of the MacArthur Project and law and neuroscience more generally. The problem is not unique to brain science but has perhaps reached its clearest expression in that context. The work has been distilled into a "knowledge brief" put out by the MacArthur Foundation Research Network on Law and Neuroscience: "Science seeks to understand general phenomena, not particular instances. Scientists typically don't attempt to infer from group or population-based data (or 'G') to a particular individual (or 'i')" (Faigman et al., 2017). But this is exactly what law (in the courtroom) must do: make a definitive decision about a specific individual/circumstance. And, of course, there is huge variation in individual brains, behavior, and circumstances. "The G2i problem is relevant for any scientific domain in which there is variability across individuals, including but not limited to neuroscientific measures" (Poldrack et al., 2018). The G2i problem is not (yet) a solvable problem, but it might be a manageable one via jury and judicial education about the nature of the necessary inferences. Current activities in the general field of "law and neuroscience" involve extensive judicial education efforts in state and federal trial and appellate courts around the country to familiarize judges with brain science basics and clearly explain the G2i problem presented by such evidence.

Finally, there is the question of whether any brain science as evidence in court has any impact at all and whether its persuasive impact too far outstrips its explanatory power. If brain science is attempted to be admitted into court, it is because the proffering party thinks it will be useful and persuasive. But the empirical evidence on whether brain science has a particularly persuasive impact on jurors (or mock jurors) and judges is mixed. Some early work on issues of admissibility hypothesized that brain images would be unduly persuasive to a legal decision-maker, assigned weight that their inferences could not logically bear (Brown and Murphy, 2010; Weisberg et al., 2008). Some experimental evidence has borne this out (Moulin et al., 2018). But in other studies, brain images appeared to have no particularly biasing effect on mock-juror subjects than non-image neuroscience testimony (Schweitzer et al., 2011; Saks et al., 2014; Saks and Schweitzer, 2011). In studies of mock jurors assigning culpability to capital defendants, sometimes neuroscience evidence was mitigating (Greene and Cahill, 2012) but in some circumstances was found to be aggravating (Saks et al., 2014). More recent work suggests a pattern of (legal) decision-making influenced by neuroscience evidence in a way that depends on one's prior beliefs and frameworks, generally consistent with a pattern of motivated reasoning (Scurich and Shniderman, 2014; Allen et al., 2019; Perricone et al., 2022). Looking back to the evidence on the impact of neuroscience on the lay public's conceptions of personhood and responsibility, there are parallels. What can we take away from these findings with respect to whether neuroscience should be admitted into the courtroom? Well, it depends, and probably on a case-by-case basis. Judges making admissibility decisions will have to weigh the evidence and the context in which it is offered, just as they do with every other piece of scientific evidence. And an opposing party will have to attempt to weaken the relevance of a piece of evidence by attacking the chains of inference or the reliability and fit for the proffered purpose, just as they do with every other type of (expert) evidence.

4.4 Neurolaw's Future: Society, Rights, and Capabilities

The future of neurolaw is varying from earlier neurolaw on two axes: first, it is taking a broader view of "law," which of course is not just criminal investigation, adjudication, and punishment/rehabilitation but also civil dispute resolution, governance via legislation and regulation (also thought of as "policy"), rights-claiming akin to international human rights, and administrative adjudication. Second, it is expanding from a focus on the individual to the group. Neurolaw's future intersects heavily with interdisciplinary work in "neuroscience and society" and "brain health," recognizing that law is a determinant of both society and of individual and public health.

This is not to say that a concern with criminal justice and courtroom applications is waning or will fade entirely. A law review symposium in 2021 on "Imagining the Future of Law and Neuroscience" resulted in an issue of articles mostly grappling with law and neuroscience as applied to issues in criminal justice. At least two articles recapitulate the hard incompatibilist argument that since there is no free will, the law is wrong about its assumptions of human agency and thus assignment of responsibility and use of retributive justification for punishment (Alces and Sapolsky, 2022; Callender, 2022). One analyzes potential "neurointerventions" as a form of criminal punishment (Shaw, 2022). Another reviews how expert testimony has largely shaped the outcome of introductions of brain evidence into court (Denno, 2022). Owen Jones helpfully organized the future potential of law and neuroscience into five categories: detecting things law cares about, individualizing developmental stages and brain states (one way of addressing the G2i problem via technological advancement), evidence-based legal reforms, legal decision-making, and brain–brain interfaces (2022). But the exciting future of neurolaw is most likely to be concerned with whether legal mechanisms – such as rights and legislation – can protect our brains and promote their individual collective thriving.

Some of this is driven by neurotechnologies and digital technologies such as AI. Nita Farahany has articulated a comprehensive theory of "cognitive liberty" to meet the age of brain-sensing wearables, implantables, and psychoactive drugs. Cognitive liberty is the right to self-determination over our brains and mental experiences and also the right to be free from interference with mental privacy and freedom of thought (Farahany, 2023). Because some workplaces are already assessing brain function, and sensors capable of decoding brain activity can be worn in earphones, Farahany urges that safeguards for mental privacy be put in place now before such technology is ubiquitous. She argues that it should be reflected in international human rights law and recognized as a strong legal and societal norm by updating our understanding of privacy to include mental privacy. Freedom of thought is defined as a freedom from interception, manipulation, and punishment of our thoughts. She also argues for the set of rights to include a right to access and use of technologies as autonomy preserving. Other scholars are debating the need for and utility of various forms of "neuro-rights" in international law (Bublitz, 2022; Fins, 2022; Ligthart et al., 2023) Advice of the "Neurorights Initiative" that advocates for a specific set of five "neurorights" (https://neurorightsfoundation.org/mission) appears to have influenced the adoption of a constitutional amendment by the Chilean legislature that purports to protect brain activity and brain data (https://neurorights-foundation.org/chile).

Cognitive liberty and neurorights respond to brain-influencing technology. But law is a broader social determinant of brain health in the context of education, caregiving, nutrition, environmental health, and every other aspect of social policy. The emerging theory of "collective cognitive capital" conceptualizes as an aggregated resource the brain health and functional capacities of a group or society. I have recently argued that the next era of neurolaw should focus on defining and growing such "collective cognitive capital" (Murphy, 2022). It is the capabilities such as attention, memory, emotional resilience, and self-control that allow us to effectively self-govern, cooperate, innovate, and thrive. The core idea is that neurolaw should look beyond just preventing crime or assigning blame and proactively consider how the legal system can foster societal conditions that allow cognitive capital to flourish at a population level. Lawmakers should draw on brain science to positively shape institutions and environments in ways that promote healthy brain development and function for everyone. This may take the form of rights-claiming by vulnerable groups (such as children) that a focus of the state should be providing them with conditions for healthy brain development – conceptual work that has begun in earnest with interdisciplinary collaboration (https://cainstitute.edu/research/an-affirmative-right-to-healthy-brain-development-for-children/).

Crucially, this expansive vision goes hand in hand with considering neurodiversity, identity (and intersecting identities), and varying conceptions of what defines optimal collective brain function. The goal isn't a conformist conception of "optimal" brains but providing everyone the conditions to freely develop their full cognitive potential while respecting differences. Collective cognitive capital isn't about maximizing the raw processing power of some "normal" brain type – it's about nurturing many diverse forms of intelligence, social cognition, and mental well-being that together drive a pluralist view of societal thriving.

So as neurolaw enters its next era, it should set its sights on this lofty ambition of providing a brain science-informed framework for legal and policy decisions aimed at boosting the shared cognitive resources that allow for sustainable, equitable human flourishing. This won't be easy – it requires grappling with deep questions of distributive justice, individuality vs. collectivism, and the tensions between liberty and equality. And, of course, neurolaw must remain vigilant to the ever-present risk of over-applications of neuroscience that violate individual rights. But the hope is that neurolaw can move beyond a reactive, individualistic focus on stopping discrete harms and towards proactively providing evidence-based legal and institutional reforms to grow the long-term mental resilience of whole populations.

To the extent that neurolaw is understood as a strain of co-production between science and society (Jasanoff, 2004), its future involves intentionally shaping the science as much as science shapes the law. Calls for such collaboration and co-production have been echoing for about a decade from within the social sciences (Rose and Abi-Rached, 2013; Fitzgerald et al., 2014; Fitzgerald and Callard, 2015). To the extent that the focus on the *law* part of neurolaw directs attention to collective problems of governance (such as establishing rights and constraining the state) and away from individual problems (such as prediction or adjudication of an individual's behavior), there is a possibility that future neurolaw can incentivize and inspire new brain/behavioral science research questions about collective decision-making, collective thriving, and build a neuroscience deeply informed by sociological theories and constructs. Already we see emerging a cutting-edge brain/behavioral science of systems and groups, defining collective forms of cognition and moving beyond methodological individualism in behavioral research (Garcia et al., 2024; Rabb and Sloman, 2024; Shteynberg, 2024). Can the impetus of neurolaw push our neuroscience methodologies away from individualism and thus our understanding of ourselves as deeply embedded, networked, and embodied beings in social, cultural, and environmental structures? The foregoing work in neurolaw has proceeded in fits and starts to incorporate the findings of neuroscience into the pursuits of law. When both disciplines aim together at promoting human thriving and collective well-being, there is cause for much hope.

References

Aharoni E, Mallett J, Vincent GM, Harenski CL, Calhoun VD, Sinnott-Armstrong W, and Kiehl KA (2014). Predictive accuracy in the neuroprediction of rearrest. *Soc. Neurosci.* 9, 332–336.

Aharoni E, Vincent GM, Harenski CL, Calhoun VD, Sinnott-Armstrong W, Gazzaniga MS, and Kiehl KA (2013). Neuroprediction of future rearrest. *Proc. Natl. Acad. Sci. USA* 110, 6223–6228.

Alces PA and Sapolsky RM (2022). Nohwere. *William Mary Law Rev.* 63, 1079–1118.

Allen CH, Vold K, Felsen G, Blumenthal-Barby JS, and Aharoni E (2019). Reconciling the opposing effects of neurobiological evidence on criminal sentencing judgments. *PLoS ONE* 14, e0210584.

Aspinwall LG, Brown TR, and Tabery J (2012). The double-edged sword: does biomechanism increase or decrease judges' sentencing of psychopaths? *Science* 337(6096), 846–849.

Barth AS (2007). A double-edged sword: the role of neuroimaging in capital sentencing cases. *Am. J. Law Med.* 33, 501–522.

Bingham T (2004). Preface. *Phil. Trans. R. Soc. B* 359, 1659.

Brooks JD (2014). What any parent knows but the supreme court misunderstands: reassessing neuroscience's role in diminished capacity jurisprudence. *New Crim. Law Rev.* 17, 442–501.

Brown T and Murphy E (2010). Through a Scanner Darkly: functional neuroimaging as evidence of a criminal defendant's past mental states. *Stan. L. Rev.* 62, 1119–1208.

Bublitz JC (2022). Novel neurorights: from nonsense to substance. *Neuroethics* 15, 7.

Callender JS (2022). Neuroscience and criminal justice: time for a "Copernican Revolution"? *William Mary Law Rev.* 63, 1119–1156.

Davis K, Flor H, Greely HT, Iannetti GD, Mackey S, Ploner M, Pustilnik A, Tracey I, Treede RD, and Wager TD (2017). Brain imaging tests for chronic pain: medical, legal and ethical issues and recommendations. *Nat. Rev. Neurol.* 13, 624–638.
Denno DW (2015). The myth of the double-edged sword: an empirical study of neuroscience evidence in criminal cases. *B.C.L. Rev.* 56, 493–551.
Denno DW (2022). How experts have dominated the neuroscience narrative in criminal cases for twelve decades: a warning for the future. *William Mary Law Rev.* 63, 1215–1296.
Dolan DD, Lee SS-J, and Cho MK (2022). Three decades of ethical, legal, and social implications research: looking back to chart a path forward. *Cell Genom.* 2, 100150.
Dumit J (2004). *Picturing Personhood: Brain Scans and Biomedical Identity.* Princeton University Press.
Faigman DL, Monahan J, and Slobogin C (2014). Group to individual (G2i) inference in scientific expert testimony. *U. Chi. L. Rev.* 81, 417–453.
Faigman DL, Wagner AD, Bonnie RJ, Casey BJ, Davis A, Hoffman MB, Jones OD, Montague R, Morse SJ, Raichle ME, Richeson JA, Scott ES, Steinberg L, Taylor-Thompson K, and Yaffe G (2017). *G2i Knowledge Brief: A Knowledge Brief of the MacArthur Foundation Research Network on Law and Neuroscience.* MacArthur Foundation Research Network on Law & Neuroscience.
Farahany NA (2012a). Incriminating thoughts. *Stanford Law Rev.* 64, 351–408.
Farahany NA (2012b). Searching secrets. *U. Pa. Law Rev.* 160, 1239–1308.
Farahany NA (2015). Neuroscience and behavioral genetics in US criminal law: an empirical analysis. *J. Law Biosci.* 2, 485–509.
Farahany NA (2023). *The Battle for Your Brain: Defending the Right to Think Freely in the Age of Neurotechnology.* St. Martin's Press.
Farwell LA and Donchin E (1991). The truth will out: interrogative polygraphy ("lie detection") with event-related brain potentials. *Psychophysiology* 28, 531–547.
Fins JJ (2022). The unintended consequences of Chile's neurorights constitutional reform: moving beyond negative rights to capabilities. *Neuroethics* 15, 26.
Fitzgerald D and Callard F (2015). Social science and neuroscience beyond interdisciplinarity: experimental entanglements. *Theory Cult. Soc.* 32, 3–32.
Fitzgerald D, Matusall S, Skewes J, and Roepstorff A (2014). What's so critical about critical neuroscience? Rethinking experiment, enacting critique. *Front. Hum. Neurosci.* 8, 365.
Garcia D, Galesic M, and Olsson H (2024). The psychology of collectives. *Perspect. Psychol. Sci.* 19, 316–319.
Garland B (2004). *Neuroscience and the Law: Brain, Mind, and the Scales of Justice.* New York: Dana Press.
Gilbert F and Dodds S (2020). Is there anything wrong with using AI implantable brain devices to prevent convicted offenders from reoffending? In: *Neurointerventions and the Law: Regulating Human Mental Capacity* (Vincent NA, Nadelhoffer T, and McCay A, eds). Oxford University Press.
Giridharadas A (2008). India's novel use of brain scans in courts is debated. *N.Y. Times*, Sept. 14, A10.
Goldstein M (1990). The decade of the brain. *Neurology* 40, 321–322.
Greely HT (2004). Prediction, litigation, privacy, and property: some possible legal and social implications of advances in neuroscience. In: *Neuroscience and the Law: Brain, Mind, and the Scales of Justice.* New York: Dana Press.
Greely HT (2015). Neuroscience, mindreading, and the courts: the example of pain. *J. Health Care L. & Pol'y* 18, 171–256.
Greely HT and Illes J (2007). Neuroscience-based lie detection: the urgent need for regulation. *Am. J. Law Med.* 33, 377–432.

Greene E and Cahill BS (2012). Effects of neuroimaging evidence on mock juror decision making. *Behav. Sci. Law* 30, 280–296.

Greene J and Cohen J (2004). For the law, neuroscience changes nothing and everything. *Phil. Trans. R. Soc. B* 359, 1775–1785.

Jasanoff S (2004). *States of Knowledge: The Co-Production of Science and Social Order*. Routledge.

Jones OD (2022). The future of law and neuroscience. *William Mary Law Rev.* 63, 1317–1343.

Jones OD and Shen FX (2012). Law and neuroscience in the United States. In: *International Neurolaw: A Comparative Analysis* (Spranger TM, ed), pp 349–362. Springer.

Jones OD and Wagner AD (2019). Law and neuroscience: progress, promise, and pitfalls. In: *The Cognitive Neurosciences*, 6th Edition (Poeppel D, Mangun GR, and Gazzaniga MS, eds), pp 1015–1025. MIT Press.

Kolber AJ (2007). Pain detection and the privacy of subjective experience. *Am. J. L. & Med.* 33, 433–456.

Kozel FA, Johnson KA, Mu Q, Grenesko EL, Laken SJ, and George MS (2005). Detecting deception using functional magnetic resonance imaging. *Biol. Psychiatry* 58, 605–613.

Kozel FA, Revell LJ, Lorberbaum JP, Shastri A, Elhai JD, Horner MD, Smith A, Nahas Z, Bohning DE, and George MS (2004). A pilot study of functional magnetic resonance imaging brain correlates of deception in healthy young men. *J. Neuropsychiatry Clin. Neurosci.* 16, 295–305.

Kulynych J (1997). Psychiatric neuroimaging evidence: a high-tech crystal ball? *Stanford Law Rev.* 49, 1249–1270.

Langleben DD, Loughead JW, Bilker WB, Ruparel K, Childress AR, Busch SI, and Gur RC (2005). Telling truth from lie in individual subjects with fast event-related fMRI. *Hum. Brain Mapp.* 26, 262–272.

Langleben DD, Schroeder L, Maldjian JA, Gur RC, McDonald S, Ragland JD, and Childress AR (2002). Brain activity during simulated deception: an event-related functional magnetic resonance study. *Neuroimage* 15, 727–732.

Lee TMC, Liu HL, Tan LH, Chan CCH, Mahankali S, Feng CM, Hou J, Fox PT, and Gao JH (2002). Lie detection by functional magnetic resonance imaging. *Hum. Brain Mapp.* 15, 157–164.

Ligthart S, Kooijmans T, Douglas T, and Meynen G (2021). Closed-loop brain devices in offender rehabilitation: autonomy, human rights, and accountability. *Camb. Q. Healthc. Ethics* 30, 669–680.

Ligthart S, van de Gronde T, Meynen G, Holland P, and Lavazza A (2023). Minding rights: mapping ethical and legal foundations of 'neurorights.' *Camb. Q. Healthc. Ethics*, 1–14.

Mackintosh N (2011). Guilty minds. *New Scientist*, Dec. 17, 26–27.

Marcus S and Dana Foundation (2002). *Neuroethics: mapping the field*. Conference proceedings, May 13–14, 2002, San Francisco, CA. Dana Press.

Maroney TA (2013). The false promise of adolescent brain science in juvenile justice. *Notre Dame Law Rev.* 85, 89–176.

Maslowsky J, Buss E, and Wray-Lake L (2024). The role (and limits) of developmental neuroscience in determining adolescents' autonomy rights: the case for reproductive and voting rights. *Dev. Cogn. Neurosci.* 69, 101435.

Moriarty JC (2008). Flickering admissibility: neuroimaging evidence in the U.S. courts. *Behav. Sci. Law* 26, 29–49.

Moriarty JC (2021). Neuroimaging evidence in US courts. In: *Law and Mind: At the Intersection of Law and the Cognitive Sciences* (Brozek B, Shen F, Vincent N, and Hage J, eds) Cambridge University Press.

Morse SJ (1996). Brain and blame. *Geo. L.J.* 84, 527–549.

Morse SJ (2004). New neuroscience, old problems. In: *Neuroscience and the Law: Brain, Mind, and the Scales of Justice*. New York: Dana Press.
Morse SJ (2006). Brain overclaim syndrome and criminal responsibility. *Ohio State J. Crim. Law* 3, 397–412.
Moulin V, Mouchet C, Pillonel T, Gkotsi G-M, Baertschi B, Gasser J, and Testé B (2018). Judges' perceptions of expert reports: the effect of neuroscience evidence. *Int. J. Law Psychiatry* 61, 22–28.
Murphy ER (2013). Paved with good intentions: sentencing alternatives from neuroscience and the policy of problem-solving courts. *Law Psychol. Rev.* 37, 100–132.
Murphy ER and Greely HT (2011). What will be the limits of neuroscience-based mindreading in the law? In: *Oxford Handbook of Neuroethics* (Illes J and Sahakian B, eds), pp 635–654. Oxford University Press.
Murphy ER and Rissman J (2020). Evidence of memory from brain data. *J. Law Biosci.* 7(1), 1–83.
Murphy ERD (2022). Collective cognitive capital. *William Mary Law Rev.* 63, 1347–1419.
O'Connor C and Joffe H (2013). How has neuroscience affected lay understandings of personhood? A review of the evidence. *Public Underst. Sci.* 22, 254–268.
Open your mind. (2002) *The Economist*.
Perricone A, Baskin-Sommers A, and Ahn W-K (2022). The effect of neuroscientific evidence on sentencing depends on how one conceives of reasons for incarceration. *PLoS ONE* 17, e0276237.
Phalen HJ, Salerno JM, and Schweitzer NJ (2021). Can neuroimaging prove pain and suffering? The influence of pain assessment techniques on legal judgments of physical versus emotional pain. *Law Hum. Behav.* 45, 393–412.
Poldrack RA, Monahan J, Imrey PB, Reyna V, Raichle ME, Faigman D, and Buckholtz JW (2018). Predicting violent behavior: what can neuroscience add? *Trends Cogn. Sci.* 22, 111–123.
Pustilnik AC (2012). Pain as fact and heuristic: how pain neuroimaging illuminates the moral dimensions of law. *Cornell Law Rev.* 98, 801–848.
Pustilnik AC (2015). Imaging brains, changing minds: how pain neuroimaging can transform the law. *Ala. Law Rev.* 66, 1099–1158.
Pustilnik AC (2018). Status and surveillance in the use of brain-based pain imaging in the law. In: *Developments in Neuroethics and Bioethics* (Buchman DZ and Davis KD, eds), pp 59–84. Academic Press.
Rabb N and Sloman SA (2024). Radical collective intelligence and the reimagining of cognitive science. *Top. Cogn. Sci.* 16, 164–174.
Raine A, Buchsbaum M, and Lacasse L (1997). Brain abnormalities in murderers indicated by positron emission tomography. *Biol. Psychiatry* 42, 495–508.
Raine A, Buchsbaum MS, Stanley J, Lottenberg S, Abel L, and Stoddard J (1994). Selective reductions in prefrontal glucose metabolism in murderers. *Biol. Psychiatry* 36, 365–373.
Rose N and Abi-Rached J (2013). *Neuro: The New Brain Sciences and the Management of the Mind*. Princeton University Press.
Rosen J (2007). The brain on the stand. *N.Y. Times Magazine*, March 11, 48–53.
Saks MJ and Schweitzer NJ (2011). Neuroimage evidence and the insanity defense. *Behav. Sci. Law* 29, 592–607.
Saks MJ, Schweitzer NJ, Aharoni E, and Kiehl KA (2014). The impact of neuroimages in the sentencing phase of capital trials. *J. Empir. Leg. Stud.* 11, 105–131.
Sapolsky RM (2004). The frontal cortex and the criminal justice system. *Phil. Trans. R. Soc. B.* 359, 1787–1796.
Sapolsky RM (2021). Neuroscience and the law. *U. St. Thomas J.L. & Pub. Pol'y* 15, 138–138.

Schweitzer NJ, Saks MJ, Murphy ER, Roskies AL, Sinnott-Armstrong W, and Gaudet LM (2011). Neuroimages as evidence in a mens rea defense: no impact. *Psychol. Public Policy Law* 17, 357–393.

Scurich N and Shniderman A (2014). The selective allure of neuroscientific explanations. *PLoS ONE* 9, e107529.

Seiden JA (2004). The criminal brain: frontal lobe dysfunction evidence in capital proceedings. *Cap. Def. J.* 16, 395–430.

Shaw E (2022). Neuroscience, criminal sentencing, and human rights. *William Mary Law Rev.* 63, 1409–1451.

Shteynberg G (2024). The psychology of collective consciousness. *J. Consum. Psychol.* 34, 678–686.

Snead OC (2008). Neuroimaging and the "complexity" of capital punishment. *N.Y.U. L. Rev.* 82, 1265–1339.

Spence SA, Farrow TF, Herford AE, Wilkinson ID, Zheng Y, and Woodruff PW (2001). Behavioural and functional anatomical correlates of deception in humans. *Neuroreport* 12, 2849–2853.

Steinberg L (2013). The influence of neuroscience on US Supreme Court decisions about adolescents' criminal culpability. *Nat. Rev. Neurosci.* 14, 513–518.

Steinberg L (2017). Adolescent brain science and juvenile justice policymaking. *Psychol. Public Policy Law* 23, 410–420.

Steinberg L, Cauffman E, Woolard J, Graham S, and Banich M (2009). Are adolescents less mature than adults? Minors' access to abortion, the juvenile death penalty, and the alleged APA "flip-flop." *Am. Psychol.* 64, 583–594.

Case Law

Frye v. United States, 293 F. 1013 (D.C. Cir. 1923)
Graham v. Florida, 560 U.S. 48 (2010)
Miller v. Alabama, 567 U.S. 460 (2012)
Roper v. Simmons, 543 U.S. 551 (2005)
United States v. Scheffer, 523 U.S. 303 (1998)
United States v. Semrau, No. 07-10074, 2010 WL 6845092 (W.D. Tenn. June 1, 2010)
United States v. Semrau, 693 F.3d 510 (6th Cir. 2012)
Wilson v. Corestaff Servs. L.P., 28 Misc. 3d 425 (N.Y. Sup. Ct. 2010)

5

Neurolaw and Psychiatry

Gerben Meynen

5.1 Introduction

Neurolaw as an academic discipline concerns itself with the study of legal implications of neuroscience and neurotechnology (Meynen, 2014). Psychiatry has a considerable intersection with the law, and this intersection is often referred to as forensic psychiatry. If neurotechnology is used in forensic psychiatry, this is, therefore, relevant from a neurolaw perspective. In this chapter, we will focus on the interface between psychiatry and *criminal law*.

Psychiatry is relevant for criminal law. Criminal law is usually not only interested in *what* defendants did but also in *why* they acted as they did and *how* they may act in the future – and how such future behavior may be influenced. In some criminal cases, the defendant's behavior appears to have been influenced by the presence of a mental illness. In such cases, psychiatrists and psychologists can be asked for their expert opinion on the reasons behind a defendant's conduct. If a mental illness is established, an offender may be considered legally insane and may be sentenced to forensic psychiatric care in order to treat the mental illness that causes danger for society (Bijlsma et al., 2019).

Neuroimaging is sometimes used in forensic psychiatry to evaluate a defendant's condition. For instance, a brain scan may reveal the presence of a tumor, or it may show signs of traumatic brain injury or dementia (Meynen, 2020). Meanwhile, the actual value of neuroscience for answering legal questions, such as those pertaining to a defendant's criminal responsibility, has been a topic of debate – and is often considered limited (Bigenwald & Chambon, 2019; Morse, 2023). In light of the continuous progress in the neurosciences in the last decades, however, it is often assumed that in the near future, the role of neuroscience, in particular neurotechnology, in forensic psychiatry may increase. Neurotechnology could possibly not only be used to make better assessments of criminal responsibility but also to evaluate the risk of recidivism, and potentially neurotechnological devices could be used to reduce the risk of harmful behavior by changing a person's brain activity (Meynen et al., 2023).

In principle, neurotechnology may enable (forensic) psychiatrists to obtain relevant information from the brain – sometimes referred to as 'brain reading' – and to modify the brain – sometimes referred to as 'brain writing' (Roelfsema et al., 2018). The first type of

DOI: 10.1201/9781032717791-5

application is related to *assessments* in forensic psychiatry, such as assessments regarding insanity and social dangerousness. The second type of application is related to *interventions*, most likely to reduce the risk of recidivism within the context of social rehabilitation. Both topics are discussed in scholarly neurolaw debates (Meynen, 2019b). Clearly, assessment and intervention can be combined in what is called 'risk management'.

This chapter will consider the current state of neurotechnology in forensic psychiatry as well as possible near-future developments. Some emphasis will be on the legal and ethical challenges entailed by the (potential) application of neurotechnology in forensic psychiatry.

Even though forensic psychiatry covers, as said, the entire intersection between psychiatry and the law, this chapter will focus on the interface with criminal law. First, we consider criminal responsibility; next we move to risk of recidivism assessment and risk management.

However, before we discuss these topics, it may be good to mention that there has been – and continues to be – a debate about neuroscience and free will. Allegedly, neuroscience would disprove free will, or at least make its existence (highly) unlikely, and therefore – as free will is considered required for responsibility – no one would be really responsible, neither morally nor legally (see, e.g., Bigenwald & Chambon, 2019; Greene & Cohen, 2004; Morse, 2007). If that would be the case – and criminal law would be revised accordingly – then, it appears, forensic psychiatric assessments of defendants regarding their criminal responsibility would no longer make sense – as no one would be criminally responsible anymore. Even though this is an ongoing debate (Nestor, 2019), I will not consider it in this Chapter; we will focus on more practical issues related to forensic psychiatry: assessment of legal insanity and dangerousness and possible neurotechnological interventions. Topics that will be discussed include reliability, safety, consent, mental and bodily integrity, and human rights.

5.2 Legal Insanity

Almost all Western countries have an insanity defense – Sweden and some states in the United States are exceptions (Simon & Ahn-Redding, 2006). The insanity defense means that even if it has been proven that a defendant committed a crime, the defendant is not held responsible because of the influence of a mental illness on his delinquent behavior at the time of the act (Meynen, 2016, 2021). The legal criteria for this defense differ considerably across jurisdictions. Often, they contain two components, which may be influenced by a mental disorder at the time of the crime. The first component concerns a person's *knowledge*, the second concerns her behavioral *control*. An example of such a legal criterion or legal standard can be found in the Model Penal Code, developed by The American Law Institute (1962) according to which

> a person is not responsible for criminal conduct if at the time of such conduct as a result of mental disease or defect he lacks substantial capacity either to appreciate

> the criminality (wrongfulness) of his conduct or to conform his conduct to the requirements of the law.

We can recognize the 'knowledge' element, phrased as *appreciation*, and the control element, phrased as one's ability to *conform* one's conduct. Other countries often have similar components, such as China and Italy (Meynen & Bijlsma, 2022). Still, arguably the most influential legal standard in the Western world, the *M'Naghten rule*, has *only one* of these components: knowledge. It reads:

> At the time of committing the act, the party accused was laboring under such a defect of reason, from disease of the mind, as not to know the nature and quality of the act he was doing; or if he did know it, that he did not know what he was doing was wrong.
>
> **M'Naghten's Case, 10 Cl. & Fin. 200, 8 Eng. Rep. 718 (H.L., 1843)**

This standard only regards the defendant's knowledge at the time of the act. So, if a person knew the nature, quality, and wrongfulness of the act, he is not legally insane, even if a mental illness prevented him from acting according to such knowledge. Many states in the US use this criterion (Packer, 2009).

As said, the criteria vary across jurisdictions. Another aspect of legal insanity that varies across jurisdictions concerns the *degrees* or *grades* of criminal responsibility. In the US, for instance, defendants are either considered responsible or legally insane (Morse, 2003), while in the Netherlands, and in many other countries, there is also an intermediate category of 'diminished' responsibility (Meynen, 2023; Simon & Ahn-Redding, 2006).

The differences between countries and jurisdictions regarding the criteria of legal insanity imply that, ultimately, there are also differences regarding the kind of psychiatric – and potentially neurotechnological – information that is of value to the legal decision-makers (judges and juries). This is a first point to keep in mind: ultimately, the law and legal decision-makers determine what is – and what is not – relevant to the case at hand, also regarding neurotechnological information.

Neuroimaging has been used in cases of criminal responsibility (De Kogel & Westgeest, 2015; Gkotsi et al., 2019). Examples of technologies that can, in principle, be used, are CT scans and MRI for structural neuroimaging and fMRI and PET scans for functional imaging (Roskies, 2013). It is often hoped that neuroimaging could, at least in some cases, add *objectivity* to forensic psychiatric assessments of defendants. As Gkotsi et al. (2019) write:

> Neurotechnologies are defended as being useful in terms of objectivity and completeness . . . and neuroscientific evidence is often perceived as more objective, reliable and 'scientific' evidence.

Adding such objectivity may be considered important because forensic psychiatric evaluation often relies, to a considerable extent, on self-report (Meynen, 2020). Surely, there may be relevant additional information in police reports, such as witness testimonies, but still, what was going on in a defendant's mind at the time of the crime is basically a subjective matter – only directly accessible to the defendant herself. Others can basically observe a person's behavior but not the reasons behind it – they are often 'hidden' in the mind. The fact that such mental content is not directly accessible to others means that information about a defendant's mental states has, at least to a substantial extent, to come from the defendant herself. However, at this point, problems can arise, as the defendant, may, for instance, have difficulty remembering what she was thinking weeks or months ago when she performed the delinquent act, she may be unwilling to share it with a psychiatrist, or she may choose to lie about it or fake certain symptoms. For instance, she may act as if she is hearing voices, feigning psychosis. Faking or malingering are well-recognized issues in forensic psychiatric assessment (Meynen, 2020; Rogers, 2012). So, adding objectivity to the assessment could be helpful.

An obvious challenge for using neuroimaging in this way is that the criteria for mental illnesses, both in the DSM-5 (American Psychiatric Association, 2013) and in the ICD-10 (World Health Organization & ebrary Inc., 1992), are almost exclusively behavioral or subjective in nature. In other words, the diagnostic criteria for mental disorders such as psychosis, bipolar disorder, antisocial personality disorder, or autism spectrum disorder usually used in (forensic) mental health care *do not* refer to specific neurobiological parameters. From this perspective, it is not immediately clear how neuroimaging could be helpful in diagnosing these disorders. Furthermore, if we look at the mentioned criteria for legal insanity – knowledge and behavioral control – it is also difficult to see how exactly neuroimaging could be helpful, as current brain imaging technologies, in principle, cannot straightforwardly determine one's legally relevant knowledge or behavioral control.

Still, even though psychiatric disorders are the most common candidates for an insanity defense and psychiatrists usually are the medical experts providing testimony, neurological disorders can be relevant to insanity evaluations as well. And neuroimaging can be very helpful in the diagnostic process for *neurological illnesses* such as cerebrovascular accidents, brain trauma, and dementia (a disorder at the intersection of neurology and psychiatry) (Darby et al., 2024). These conditions may be highly relevant to questions of insanity or diminished criminal responsibility.

Still, even though, depending on the case, neurobiological measurements may provide solid data about certain conditions, there may still be ample room for interpretation. As Gkotsi et al. (2019) write:

> it is generally observed that the very same neurobiological evidence is analyzed and interpreted in different – sometimes diametrically opposed – ways from the defense and prosecution experts.
>
> **(p. 7)**

Let us consider a case in which neurobiological assessment yielded information that the court took into account: a Dutch case of attempted extortion – which attracted a lot of media attention (Court reference: ECLI:NL:RBMNE:2015:4866). Several behavioral experts, among them a neurologist, evaluated the defendant, an elderly man. Frontotemporal dementia, behavioral subtype, was diagnosed, with profound impact on the defendant's mental capacities, influencing his delinquent behavior in such a way that his responsibility was considered strongly diminished. However, in the media, doubts had been raised that the defendant may have been faking (Thijssen, May 19, 2015). Eventually, on the Judicial System Netherlands website, it was stated, based on the court ruling: "It is very unlikely that the defendant can fake his brain disorder, this is evident from the MRI scan and PET scan that were shown during the session and assessment by the behavioral experts" (Rechtspraak.nl; see also Meynen, 2021). Apparently, for the court, in this case neuroimaging supported the other findings by the behavioral experts, providing data that *could not be faked*.

Obviously, the mere fact that neuroimaging or other neurobiological data are used in court cases and that they may be taken into account by the judge or jury in a way that impacts the verdict regarding legal insanity does not necessarily imply that this happened for the right reasons (Meynen, 2020) (Morse, 2023). In other words, not every impact neurotechnology has on legal decision-making is necessarily positive; neuroscientific data could, for instance, be overinterpreted or misinterpreted. Yet the same is clearly true for other types of biological evidence, most notably DNA evidence: the correct interpretation of the findings is always a crucial issue. Still, many will agree that DNA evidence is often of enormous value to the criminal justice system (Meynen, 2016). (At the same time, while there is a lot of expertise and standardized procedures regarding DNA evidence in criminal cases, the use of neuroscience with respect to criminal responsibility questions is less standardized.)

Concerning the interpretation of neuroimaging findings regarding legal insanity, *neuropsychological* evaluation may be valuable (Meynen, 2021). For instance, if there is a match between MRI findings and observed neuropsychological deficits, this may support the hypothesis that the MRI findings reflect changes that actually impact a person's mental functioning.

Let us look at a Dutch case example of a defendant accused of fornication with an approximately nine-year-old neighbor girl that took place five times, as reported by De Kogel and Westgeest, 2015 (the quotes that follow are taken from their publication). The neurologist who evaluated the defendant diagnosed beginning fronto-subcortical dementia within the context of Parkinson's disease. He reports about several control-related issues:

> Directly related to the fronto-subcortical dementia is, in addition to the characteristic cognitive impairments, impulsivity, which occurs particularly in complex situations in which the defendant lacks an overview. As a result the defendant

> was unable to control his impulses. He acted reflexively without overseeing the consequences.
>
> **(De Kogel & Westgeest, 2015, based on Court case ECLI:NL:RBALK:2008:BK5962)**

And the neurologist adds:

> In addition he was as a result of his cognitive impairment not able to interrupt his behavior once started.

Apart from these control or impulsivity related problems, there are also more knowledge- or appreciation-related problems: "The defendant lacks the capacity for self-reflection, which prevents him from relating his actions to an appropriate framework of norms and values" (De Kogel & Westgeest, 2015, based on Court case ECLI:NL:RBALK:2008:BK5962).

In line with the advice by the experts, the Court decides that the defendant's criminal responsibility was diminished. The case makes clear that neurological examination may not only reveal a brain illness (dementia) but also problems regarding both knowledge/appreciation (broadly construed) and behavioral control – two aspects of human functioning that may be relevant in view of legal criteria for insanity.

Regarding neuroimaging, neurological, and neuropsychological evaluation, it is important to realize that these assessments take place here and now, while the crime may have been committed weeks, months, or sometimes even years before. The assessment of insanity is, in other words, an evaluation *in retrospect* (Meynen, 2021). Therefore, in principle, brain changes detected using neuroimaging in an insanity evaluation may or may not have been present at the time of the crime. In other words, one needs to explain whether today's findings are likely to have been present at the time of the crime.

Interestingly, some neurologists recently advocated "the development of 'forensic neurology', a subspecialty of neurology focused on using neurological clinical and scientific expertise to address legal questions for the criminal justice system" (Darby et al., 2024). Among others, they point to the fact that "Neurological evidence is increasingly used in criminal court cases, with evidence being introduced in up to 12% of all murder trials and 25% of all capital cases in the USA over the past decade". The establishment of such a neurological subspecialty could, they write, create more awareness of the importance of brain functioning as related to criminal behavior. More specifically, it could be beneficial to the quality of assessments of legal insanity. One of the further benefits they mention would be more research on the relationship between "brain and crime" (see also Swaab & Meynen, 2023).

In sum, at this point in time, neuroimaging plays a limited role in assessments of criminal responsibility, but still, it may be of value in individual cases, particularly in those where neurological conditions are suspected. And we cannot exclude the possibility that scientific advances regarding the neurobiological correlates of mental illness – and specific mental deficits or incapacities – will lead to a broader applicability of neuroimaging technologies within the context of insanity assessments.

5.3 Risk of Recidivism

While assessments of legal insanity are backward looking in time, assessments of risk of recidivism concern the future. Since mental illnesses may play a role in the risk of recidivism, psychiatrists may be asked in court cases to evaluate such a risk.

Neurotechnology may also be used in such a forward-looking way (Gkotsi, 2021) (Meynen et al., 2023) (Geukes et al., 2024). For instance, in the attempted extortion case mentioned in the previous section, the frontotemporal dementia with which the defendant was diagnosed was not only taken into account regarding the question of legal insanity but also concerning the question of risk of recidivism. The risk of recidivism in this case was considered very low, one consideration being that "that frontotemporal dementia is a progressive disease and that the defendant's overall functioning will only continue to deteriorate". In other words, the disease may soon compromise the cognitive abilities needed for the kind of crime he committed; the defend will be too ill for committing further crimes. The defendant was sent to a mental health institution specialized in frontotemporal dementia (ECLI:NL:RBMNE:2015:4866, own translation in quote).

Interestingly, as Khalid et al. state, studies suggest a "double edged sword" regarding the use of neuroscience in the courtroom as related to backward- and forward-looking questions of the law:

> As in the United States, Canadian courts most frequently consider neuroscientific evidence to inform criminal sentencing, where it can have both mitigating and aggravating effects, through a "double-edged sword" effect, whereby such evidence supports diminished culpability while simultaneously increasing perceptions of risk and future dangerousness.
>
> **(Khalid et al., 2024)**

While this does not appear to be true for the attempted extorsion case, it might be true for the fornication case. In that case, the psychologist reported

> that the dementia is in an early phase and that the behavior of the defendant shows not a general disinhibition yet. She reasons that there is a realistic risk of criminal recidivism.
>
> **(De Kogel & Westgeest, 2015)**

In both dementia cases, a defendant's *neuropsychopathological condition* is relevant for the risk of recidivism. Interestingly, however, neuroimaging studies have also been used to assess risk of recidivism in what can be considered a *direct* manner, so not by providing support for a particular mental disorder but by the analysis of (proxies of) brain activity itself – and relating the measurements to risk of recidivism. An example of such 'direct' assessment of risk of recidivism is an often-cited study by Aharoni et al. who reported that, using fMRI, it was possible to predict who would be rearrested in a period of four years after release (2013). The prediction was not perfect but significantly better than chance. While this study was performed in male offenders, the same research group recently reported that it was also possible to predict reoffending based on fMRI data in female offenders (Allen et al., 2022). In a Swedish study, neuroimaging data were *combined* with more traditional risk factors, such as criminal history, in order to predict recidivism in a forensic psychiatric population using an algorithm (Delfin et al., 2019). It turned out that the prediction of recidivism improved when neuro-data were added. One could imagine that, in the future, data from different sources, such as criminal history, neuroimaging, psychological tests, and clinical data, can be combined and analyzed by an algorithm in order to make an accurate risk assessment.

As evaluations of risk of recidivism regard the future, it is, in principle, possible to *intervene* based on such an assessment. For instance, if a mental disorder is a factor contributing to the risk of recidivism, the defendant may be sent to a forensic mental health facility. In such a facility, forensic patients may be treated, among others, with psychotherapy and medication. In fact, an array of rehabilitative interventions is, in principle, available in criminal justice, ranging from social to psychological and medical interventions (McNeill, 2012).

Perhaps, in the near future, it may also become possible to use neurotechnological devices within the context of rehabilitation (Geukes et al., 2024). In a paper published in 2015, Fuss and colleagues considered the future possibility of using deep brain stimulation (DBS) in sex offenders. Drawing on earlier research with psychosurgery, they (preliminarily) suggest the ventromedial thalamus as candidate location for DBS. Recently, a research group in Rotterdam has reported that application of transcranial direct current stimulation (tDCS) in a forensic psychiatric population resulted in lower self-reported aggression (Sergiou et al., 2021). More recently, German researchers observed that tDCS used in violent offenders increased prefrontal activity and improved decision-making in violent offenders (an effect not observed in the non-violent control group) (Kuhn et al., 2024). To my knowledge, meanwhile, neuromodulation – tDCS, DBS, or transcranial magnetic stimulation (TMS) – is not (yet) used in forensic practice for rehabilitation purposes, but the cited research suggests that in the near future such a possibility may be considered.

As mentioned, basically, there are two kinds of neurotechnologies, those that obtain information from the brain and those that intervene in the brain. But these features can be combined in so-called closed-loop devices (Ligthart et al., 2023b). On the one hand, such a device 'reads' brain activity, while on the other, it can modify its output based on such information. Currently, closed-loop devices for neurological and mental illnesses are under development (Soleimani et al., 2023). In a forensic psychiatric context, one could imagine that the device detects brain activity that is related to a high risk of imminent aggression. At that point, the device may, based on an algorithm, generate neuromodulatory output that counters such brain activity, thus reducing the risk of an aggressive outburst (Ligthart et al., 2021). The device will then continue to monitor the brain activity, and, thus, it is able to check the effects of its own intervention – and, if necessary, it may intervene again. Such devices are considered 'closed loop' because the response to the detected brain activity does not require human intervention but is generated 'autonomously' by the device (Ligthart et al., 2021).

In sum, in the future, neuroimaging may be valuable not just regarding assessments of legal insanity but also concerning questions pertaining to the risk of recidivism. Neuromodulation could also provide a novel treatment avenue for forensic psychiatry; a recent study shows reduction of self-reported aggression using tDCS. A new development regarding neurotechnologies is closed-loop neuromodulation in which the device detects brain activity and – using an algorithm – modulates its output based on the input it receives. In principle, at least from a technological perspective, in the future, such a device could combine elements of risk assessment and risk reduction.

5.4 Ethical and Legal Concerns

Neurolaw is not just about what is – or might soon be – technologically possible at the intersection of law and neuroscience. Much of the neurolaw research regards *ethical and legal* implications – and concerns – with respect to the (potential) use of neuroscience in the justice system. There are several ethical and legal worries that have been expressed regarding the application of neurotechnology in forensic psychiatry (Meynen et al., 2023). We will consider some of them subsequently.

First of all, concerns have been raised with respect to the reliability and safety of a technology. For example, regarding neuroimaging, it is crucial that the technique be reliable so the right (diagnostic) conclusions can be drawn. In fact, there is an ongoing debate about the value of neuroimaging for evaluations of criminal responsibility. Some, among them presumably most prominently Morse, have expressed concerns that neuro-data are overvalued in the courtroom. Morse even coined the "brain overclaim syndrome", pointing to the, in his view, very limited – and often overstated – value of neuroscience regarding matters related to criminal responsibility (2006). Also recently, he pointed to its limitations regarding the questions the law is ultimately interested in, in particular criminal responsibility (Morse, 2023). Clearly, it is important to be aware of the limitations of the neuroscience and neurotechnologies used, but, on the other hand, it is also important not to ignore potentially valuable technologies that may contribute to doing justice in criminal law, even if it is only in a limited number of cases, as in each case, society aspires to do justice.

Within this context, it is interesting that, from a neurological perspective, Darby (2023) suggests the following approach for forensic experts regarding assessments of the relationship between neurological illnesses and criminal actions:

> Forensic expert witnesses in such cases must make three determinations:
>
> **(1)** Does the subject have a valid neurological disease?
> **(2)** How did this neurological disease affect behavior in this subject?
> **(3)** In what way did this behavioral change contribute to the crime in question?
>
> **(p. 51)**

And they add:

> A firmly established neurological diagnosis is critical in all such cases. Neuroimaging findings in the absence of an established neurological diagnosis have a high risk for being false positive findings and should be interpreted with extreme caution.
>
> **(p. 51)**

Even though a cautious neurological approach is helpful, it is also important to realize that criminal justice, at least in principle, may have *its own standards* for admissibility of evidence produced by particular techniques (Geukes et al., 2024). In other words, the standards of reliability used in the context of healthcare need not be the same as those in criminal justice. For instance, in the US, there are different standards or criteria that determine whether certain techniques can be used in criminal cases: *Daubert* and *Frye*. Without going into the details of these criteria, it may be that the same technology is admissible in a *Daubert* jurisdiction, while it is not in a *Frye* jurisdiction. So, it is ultimately up to the legal system to put value on certain findings. Nevertheless, clearly, psychiatrists should ensure that their statements, conclusions, and advice are well supported by the findings.

It is also relevant that Geukes et al. (2024) write that "courts consider the totality of the available evidence and not just a single source. So, even when neuroscientific evidence alone would not be fully convincing, *together with* other sources of information the legal threshold of proof may still be met". In other words, while certain neurobiological evidence in itself may not be sufficient for a particular legal outcome, together with other findings, it may. In that sense, the question is one of the 'added value' of neurodata (Meynen, 2016). Still, it is crucial that the technologies used in criminal proceedings be reliable and that their level of reliability be known and communicated to the parties involved in the case, most importantly the legal decision-makers.

As it is likely that the technologies that will be applied in the future – *if* they will be applied – use some form of artificial intelligence (AI) or algorithms, concerns regarding AI are relevant here as well. Perhaps most prominently the risk of *bias* and the fact that AI may lack *transparency* (which may make it more difficult to detect the nature of the bias) have to be taken into account (Tortora et al., 2020). As the ethics of AI is a topic of its own, we will not consider it further in this chapter.

While reliability is a prime concern regarding neuroimaging (so, obtaining data from the brain and making inferences based on them), effectiveness and safety are crucial issues regarding brain interventions. In fact, history has shown that the introduction of brain interventions in psychiatry may lead to highly problematic practices. Examples are lobotomies used for the 'treatment' of mental illnesses in the 1930s–1950s (Faria, 2013) and electroconvulsive therapy (ECT) as it has been applied, mainly in the 1950s–1970s (Reisner, 2003) – the latter forcefully depicted in the movie *One Flew Over the Cuckoo's Nest*. Meanwhile, the fact that neurointerventions have led to such problematic applications in psychiatry does not in itself imply that neurointerventions will always lead to such undesirable consequences (Ryberg, 2020). In fact, while the use of ECT was problematic some decades ago, at this moment it is applied in a way that may clearly benefit patients with severe mental illness (Leiknes et al., 2012), such as depression and catatonia, particularly in the elderly (Chatham et al., 2022). As mentioned, only very limited research has been done regarding brain interventions (such as tDCS) in the realm of criminal justice and forensic psychiatry. And it is fair to say that, even though some studies have been performed, their real-world effectiveness has yet to be established.

The safety of interventions, including their side effects, is crucial. There are considerable differences regarding the invasiveness and potential complications and side effects of various types of neuromodulation technologies, such as tDCS, TMS, and DBS. While DBS requires brain surgery (with possible surgical and anesthetic complications) TMS and tDCS do not; they are considered 'non-invasive'. Also, the side effects may differ between these technologies; for instance, tDCS appears to have only limited side effects (Brunoni et al., 2011) (Nikolin et al., 2018). Notably, meanwhile, there is no general threshold for the side effects that are acceptable. In the treatment of more serious conditions in medicine, generally, more severe (possible) complications and side effects are accepted. Still, they are a concern, and they should be known and clearly communicated to the patient or, more generally, to the person receiving the intervention.

Suppose now that in the near future one of the technologies considered for actual use in forensic psychiatry would be deemed sufficiently reliable, effective, and safe in terms of complications and side effects – would that in itself be enough for responsible application? The answer is: no, not necessarily. In the literature, several ethical and legal concerns have been raised that may be relevant *even if* the technology as such is reliable, effective and safe. We will discuss some of them here briefly.

A first concern regards *privacy* or *confidentiality* of neuroimaging data. If it would become possible to 'read' a person's mind in a forensic psychiatric setting, such as within the context of a defendant's legal insanity assessment, confidentiality is crucial. Clearly, in general, in medicine, confidentiality is central (Beauchamp & Childress, 2019). But it is good to realize that the kind of information forensic psychiatrists are interested in may be considered particularly private, as it pertains to someone's desires, delusions, fears, frustrations, fantasies, traumas, and aspirations – things one may not have shared with anyone (see also Meynen, 2020). It would be of utmost importance that information retrieved in this way by technological devices be treated confidentially, making sure that only those involved receive such information.

A second concern regards the voluntariness of the consent to neurotechnological interventions. Consider the following scenario: a forensic psychiatric patient, convicted for a sex offence, whose recidivism risk remains high even after many years of treatment in a forensic hospital. Suppose now that this patient is offered some form of neuromodulation by a psychiatrist, for instance, a closed-loop DBS device that is able to detect the brain correlates of sexual arousal and other brain activity patterns that may increase the likelihood of reoffending – and that, in such circumstances, can effectively intervene to reduce that likelihood to 'safe' levels. For this patient, accepting such an offer may well appear to be the *only way* to get out of the forensic mental health facility. The question that may arise then is whether a forensic psychiatric patient's consent to such an invasive intervention can be considered voluntary, or would it be more like, in the words of the Godfather, an "offer one cannot refuse" (Meynen, 2019b)? Some may argue that this type of offer is a "coercive" offer (Feinberg, 1986; see also Pugh, 2020). In other words, the voluntariness of a patient's consent to such an offer may be an issue of debate. It was also considered, at least in some formulations of the problem, by Fuss et al. when they wrote about the (still only theoretical) possibility of offering DBS to sex offenders (see section 3):

> The motivation for DBS treatment should not be connected to any hopes/promises of prematurely leaving prison or a forensic treatment facility as in the 1970s when most of the imprisoned sexual offenders were released after the operation. The boundary between treatment and punishment must be very clear.
>
> **(2015)**

Another concern regards the fact that a forensic psychiatric patient's consent may be compromised by a serious mental illness. Consequently, the patient may not be able to make an autonomous – competent – choice about the treatment offer. It is good to be aware of this possibility. At the same time, it is equally important to enable psychiatric patients, also if they are involuntarily admitted to a forensic psychiatric hospital, to make *their own* choices and not to just assume incompetency merely based on the presence of a certain mental disorder (Meynen & Widdershoven, 2012; Ligthart et al., 2023a). In general, until proven otherwise, patients – including those with mental illness – have to be considered competent. And if they would be considered to lack competency, health care workers should aim to restore it. Nevertheless, it is clearly important to pay attention to the issue of incompetency.

So far, we have looked at neurointerventions being *offered* to forensic psychiatric patients to which the patient competently and freely may – or may not – consent. Suppose now that valid consent *has not been obtained* and that one would want to use such technologies in a non-consensual manner (Ryberg, 2021). In such a case, the legal rights to *mental and bodily integrity* are clearly relevant (Meynen et al., 2023). Both are protected by human rights/legal frameworks. According to Article 3(1) of the Charter of Fundamental Rights of the European Union (CFR), for instance, "Everyone has the right to respect for his or her physical and mental integrity." Still, the exact protection provided by these human rights frameworks against neurotechnologies is not yet elucidated and a topic of scholarly debate. One reason the protection is unclear is that there has not yet been a human rights court case about such technologies. In general, the protection provided by various human rights frameworks against emerging neurotechnologies is currently an important field of study (Ligthart et al., 2023b). This field of inquiry, meanwhile, is much broader than forensic psychiatry (covering all kinds of implications of neurotechnologies in society), and that is why we will not consider it further in this chapter.

This chapter has focused on (possible future) uses of neurotechnologies at the intersection of psychiatry and criminal law. However, it is also imaginable that these technologies may be used in other areas at the intersection of law and psychiatry, for instance, in mental health law and, particularly, concerning involuntary admissions to non-forensic mental health institutions. Also in these areas, it may be informative to obtain information about a person's mind or brain in order to assess, for instance, dangerousness to oneself or to others. Additionally, it may be that in the future, neurotechnology could play a valuable role in assessments of competency, which may also be highly important from a legal perspective (see also Meynen, 2019a). Nevertheless, so far, discussions in the literature have often revolved around assessments of criminal responsibility and possible interventions in forensic psychiatry/criminal justice, on which we have focused in this chapter.

5.5 Conclusions

Neurolaw as an academic discipline concerns itself with actual and possible implications and applications of neuroscience and neurotechnology for the law. The use of neurotechnologies in forensic psychiatry is one of the areas of study.

Neurotechnologies have been used in forensic psychiatry, but, so far, their use has been very limited, as it is mainly restricted to neuroimaging in assessments of criminal responsibility. Worries have been expressed that too much weight may be attached to neuroimaging when answering ultimately legal questions about responsibility for one's actions.

Meanwhile, there is little doubt that, at least in some cases, neuroimaging can be helpful establishing a diagnosis in a defendant, for instance, in cases of stroke or dementia. However, even in such cases, it is important to realize that a brain change does not immediately imply reduced or absent responsibility. Careful reasoning based on state-of-the-art psychiatric examination is essential, and, if required, neurological and neuropsychological assessment should be performed as well. In any case, neurotechnologies used in forensic practice should be sufficiently reliable and safe – and their limitations should be known and communicated to the legal decision-makers.

In neurolaw, not only the current state of neurotechnology is considered, but potential near future developments are also a topic of research. One of them concerns the possible future application of neurointerventions for rehabilitative purposes. This possibility raises important questions concerning the technology's effectiveness and safety but also regarding the patient's competence and freedom when choosing such interventions, as well as the patient's mental and bodily integrity. As far as neuro-devices obtain information from the brain, privacy is a central concern.

If, in the coming years, more or better neurotechnological applications will become available in forensic psychiatry, it is crucial that they be used responsibly. It is a core task for neurolaw to elucidate how this can be done.

Acknowledgments

This publication is part of the project Law and Ethics of Neurotechnology in Criminal Justice (LENC) (with project number VI.C.201.067) of the research program Vici, which is financed by the Dutch Research Council (NWO).

References

Aharoni, E., Vincent, G. M., Harenski, C. L., Calhoun, V. D., Sinnott-Armstrong, W., Gazzaniga, M. S., & Kiehl, K. A. (2013). Neuroprediction of future rearrest. *Proc Natl Acad Sci U S A*, *110*(15), 6223–6228. https://doi.org/10.1073/pnas.1219302110

Allen, C. H., Aharoni, E., Gullapalli, A. R., Edwards, B. G., Harenski, C. L., Harenski, K. A., & Kiehl, K. A. (2022). Hemodynamic activity in the limbic system predicts reoffending in women. *Neuroimage Clin*, *36*, 103238. https://doi.org/10.1016/j.nicl.2022.103238

American Psychiatric Association. (2013). *Diagnostic and statistical manual of mental disorders: DSM-5* (5th ed.). American Psychiatric Association.

Beauchamp, T. L., & Childress, J. F. (2019). *Principles of biomedical ethics* (8th ed.). Oxford University Press.

Bigenwald, A., & Chambon, V. (2019). Criminal responsibility and neuroscience: No revolution yet. *Front Psychol*, *10*, 1406. https://doi.org/10.3389/fpsyg.2019.01406

Bijlsma, J., Kooijmans, T., de Jong, F., & Meynen, G. (2019). Legal insanity and risk: An international perspective on the justification of indeterminate preventive commitment. *Int J Law Psychiatry*, *66*, 101462. https://doi.org/10.1016/j.ijlp.2019.101462

Brunoni, A. R., Amadera, J., Berbel, B., Volz, M. S., Rizzerio, B. G., & Fregni, F. (2011). A systematic review on reporting and assessment of adverse effects associated with transcranial direct current stimulation. *Int J Neuropsychopharmacol*, *14*(8), 1133–1145. https://doi.org/10.1017/S1461145710001690

Chatham, A. N., Shafi, H., & Hermida, A. P. (2022). The use of ECT in the elderly-looking beyond depression. *Curr Psychiatry Rep*, *24*(9), 451–461. https://doi.org/10.1007/s11920-022-01353-0

Darby, R. R. (2023). Network localization of antisocial behavior in neurological patients: Evidence and implications. In H. Swaab & G. Meynen (Eds.), *Brain and crime* (Vol. 197). Elsevier.

Darby, R. R., Considine, C., Weinstock, R., & Darby, W. C. (2024). Forensic neurology: A distinct subspecialty at the intersection of neurology, neuroscience and law. *Nat Rev Neurol*. https://doi.org/10.1038/s41582-023-00920-0

De Kogel, C. H., & Westgeest, E. J. (2015). Neuroscientific and behavioral genetic information in criminal cases in the Netherlands. *J Law Biosci*, *2*(3), 580–605. https://doi.org/10.1093/jlb/lsv024

Delfin, C., Krona, H., Andiné, P., Ryding, E., Wallinius, M., & Hofvander, B. (2019). Prediction of recidivism in a long-term follow-up of forensic psychiatric patients: Incremental effects of neuroimaging data. *PLoS One*, *14*(5), e0217127. https://doi.org/10.1371/journal.pone.0217127

Faria, M. A. (2013). Violence, mental illness, and the brain – a brief history of psychosurgery: Part 1 – from trephination to lobotomy. *Surg Neurol Int*, *4*, 49. https://doi.org/10.4103/2152–7806.110146

Feinberg, J. (1986). *The moral limits of the criminal law. Volume 3, Harm to self*. Oxford University Press.

Fuss, J., Auer, M. K., Biedermann, S. V., Briken, P., & Hacke, W. (2015). Deep brain stimulation to reduce sexual drive. *J Psychiatry Neurosci*, *40*(3), 150003.

Geukes, S. H., Bijlsma, J., Meynen, G., Raemaekers, M. A. H., Ramsey, N. F., Simon Thomas, M. A., . . . Vansteensel, M. J. (2024). Neurotechnology in criminal justice: Key points for neuroscientists and engineers. *J Neural Eng*, *21*(1). https://doi.org/10.1088/1741-2552/ad1785

Gkotsi, G. (2021). Neuroscience and dangerousness evaluations: The effect of neuroscience evidence on judges. Findings from a focus group study In S. Ligthart, D. Van Toor, T. Kooijmans, T. Douglas, & G. Meynen (Eds.), *Neurolaw. Advances in neuroscience, justice & security*. Palgrave Macmillan.

Gkotsi, G. M., Gasser, J., & Moulin, V. (2019). Neuroimaging in criminal trials and the role of psychiatrists expert witnesses: A case study. *Int J Law Psychiatry*, *65*, 101359.

Greene, J., & Cohen, J. (2004). For the law, neuroscience changes nothing and everything. *Philos Trans R Soc Lond B Biol Sci*, *359*(1451), 1775–1785. https://doi.org/10.1098/rstb.2004.1546

Khalid, Z., Lee, R., & Wall, B. W. (2024). The use of neurobiological evidence in sentencing mitigation. *Behav Sci Law*. https://doi.org/10.1002/bsl.2645

Kuhn, L., Choy, O., Keller, L., Habel, U., & Wagels, L. (2024). Prefrontal tDCS modulates risk-taking in male violent offenders. *Sci Rep*, *14*(1), 10087. https://doi.org/10.1038/s41598-024-60795-z

Leiknes, K. A., Jarosh-von Schweder, L., & Høie, B. (2012). Contemporary use and practice of electroconvulsive therapy worldwide. *Brain Behav*, *2*(3), 283–344. https://doi.org/10.1002/brb3.37

Ligthart, S., Dore-Horgan, E., & Meynen, G. (2023a). The various faces of vulnerability: Offering neurointerventions to criminal offenders. *J Law Biosci*, *10*(1), lsad009. https://doi.org/10.1093/jlb/lsad009

Ligthart, S., Kooijmans, T., Douglas, T., & Meynen, G. (2021). Closed-loop brain devices in offender rehabilitation: Autonomy, human rights, and accountability. *Camb Q Healthc Ethics*, *30*(4), 669–680. https://doi.org/10.1017/S0963180121000141

Ligthart, S., Ienca, M., Meynen, G., Molnar-Gabor, F., Andorno, R., Bublitz, C., . . . Kellmeyer, P. (2023b). Minding rights: Mapping ethical and legal foundations of 'neurorights'. *Camb Q Healthc Ethics*, 1–21. https://doi.org/10.1017/S0963180123000245

McNeill, F. (2012). Four forms of 'offender' rehabilitation: Towards an interdisciplinary perspective. *Leg Criminol Psychol*, *17*(1), 18–36.

Meynen, G. (2014). Neurolaw: Neuroscience, ethics, and law. Review essay. *Ethical Theory Moral Pract*. https://doi.org/10.1007/s10677-014-9501-4

Meynen, G. (2016). *Legal insanity. Explorations in psychiatry, law, and ethics*. Springer.

Meynen, G. (2019a). Ethical issues to consider before introducing neurotechnological thought apprehension in psychiatry. *AJOB Neurosci*, *10*(1), 5–14. https://doi.org/10.1080/21507740.2019.1595772

Meynen, G. (2019b). Forensic psychiatry and neurolaw: Description, developments, and debates. *Int J Law Psychiatry*, *65*. https://doi.org/10.1016/j.ijlp.2018.04.005

Meynen, G. (2020). Neuroscience-based psychiatric assessments of criminal responsibility: Beyond self-report? *Camb Q Healthc Ethics*, *29*(3), 446–458. https://doi.org/10.1017/S0963180120000195

Meynen, G. (2021). The insanity defense. In B. Brożek, J. Hage, & N. A. Vincent (Eds.), *Law and mind: A survey of the law and the cognitive sciences*. Cambridge University Press.

Meynen, G. (2023). Legal insanity in the Netherlands: Regulations and reflections. In W. Brookbanks & R. Mackay (Eds.), *The insanity defence. International and comparative perspectives* (pp. 274–294). Oxford University Press.

Meynen, G., & Bijlsma, J. (2022). Culpability and accountability: The insanity defense. In C. Garofalo & J. J. Sijtsema (Eds.), *Clinical forensic psychology*. Palgrave MacMillan.

Meynen, G., Van de Pol, N., Tesink, V., & Ligthart, S. (2023). Neurotechnology to reduce recidivism: Ethical and legal challenges. In H. Swaab & G. Meynen (Eds.), *Brain and crime*. Elsevier.

Meynen, G., & Widdershoven, G. (2012). Competence in health care: An abilities-based versus a pathology-based approach. *Clin Ethics*, *7*(1), 39–44.

Morse, S. J. (2003). Diminished rationality, diminished responsibility. *Ohio State J Crim Law*, 289–308.

Morse, S. J. (2006). Brain overclaim syndrome and criminal responsibility: A diagnostic note. *Ohio State J Crim Law*, *3*, 397–312.

Morse, S. J. (2007). The non-problem of free will in forensic psychiatry and psychology. *Behav Sci Law*, *25*(2), 203–220. https://doi.org/10.1002/bsl.744

Morse, S. J. (2023). Neurolaw: Challenges and limits (Chapter 16). In H. Swaab & G. Meynen (Eds.), *Brain and crime*. Elsevier.

Nestor, P. G. (2019). In defense of free will: Neuroscience and criminal responsibility. *Int J Law Psychiatry*, *65*, 101344. https://doi.org/10.1016/j.ijlp.2018.04.004

Nikolin, S., Huggins, C., Martin, D., Alonzo, A., & Loo, C. K. (2018). Safety of repeated sessions of transcranial direct current stimulation: A systematic review. *Brain Stimul*, *11*(2), 278–288. https://doi.org/10.1016/j.brs.2017.10.020

Packer, I. K. (2009). *Evaluation of criminal responsibility*. Oxford University Press.

Pugh, J. (2020). *Autonomy, rationality, and contemporary bioethics*. Oxford University Press.

Reisner, A. D. (2003). The electroconvulsive therapy controversy: Evidence and ethics. *Neuropsychol Rev*, *13*(4), 199–219. https://doi.org/10.1023/b:nerv.0000009484.76564.58

Roelfsema, P. R., Denys, D., & Klink, P. C. (2018). Mind reading and writing: The future of neurotechnology. *Trends Cogn Sci*, *22*(7), 598–610. https://doi.org/10.1016/j.tics.2018.04.001

Rogers, R. (Ed.). (2012). *Clinical assessment of malingering and deception*. The Guilford Press.

Roskies, A. L. (2013). Brain imaging techniques. In S. J. Morse & A. L. Roskies (Eds.), *A primer on criminal law and neuroscience. A contribution of the Law and Neuroscience Project, supported by the MacArthur Foundation*. Oxford University Press.

Ryberg, J. (2020). *Neurointerventions, crime, and punishment: Ethical considerations*. Oxford University Press.

Ryberg, J. (2021). Neurointerventions and crime prevention: On ideal and non-ideal considerations. In S. Ligthart, D. Van Toor, T. Kooijmans, T. Douglas, & G. Meynen (Eds.), *Neurolaw: Advances in neuroscience, justice & security*. Palgrave Macmillan.

Sergiou, C. S., Santarnecchi, E., Romanella, S. M., Wieser, M. J., Franken, I. H. A., Rassin, E. G. C., & van Dongen, J. D. M. (2021). Transcranial direct current stimulation targeting the ventromedial prefrontal cortex reduces reactive aggression and modulates electrophysiological responses in a forensic population. *Biol Psychiatry Cogn Neurosci Neuroimaging*. https://doi.org/10.1016/j.bpsc.2021.05.007

Simon, R. J., & Ahn-Redding, H. (2006). *The insanity defense, the world over*. Lexington Books.

Soleimani, G., Nitsche, M. A., Bergmann, T. O., Towhidkhah, F., Violante, I. R., Lorenz, R., . . . Ekhtiari, H. (2023). Closing the loop between brain and electrical stimulation: towards precision neuromodulation treatments. *Transl Psychiatry*, *13*(1), 279. https://doi.org/10.1038/s41398-023-02565-5

Swaab, H., & Meynen, G. (2023). Introduction: On brain and crime. In H. Swaab & G. Meynen (Eds.), *Brain and crime*. Elsevier.

Thijssen, W. (May 19, 2015). Vraagtekens bij dementie afperser familie de Mol. *De Volkskrant*.

Tortora, L., Meynen, G., Bijlsma, J., Tronci, E., & Ferracuti, S. (2020). Neuroprediction and A.I. in forensic psychiatry and criminal justice: A neurolaw perspective. *Front Psychol*, *11*, 220. https://doi.org/10.3389/fpsyg.2020.00220

World Health Organization & ebrary Inc. (1992). *The ICD-10 classification of mental and behavioural disorders clinical descriptions and diagnostic guidelines*. World Health Organization.

6
Stochastic Determinism and Criminal Law

Sukumar Vijayaraghavan

Belief in free will means never having to acknowledge your own great good fortune or recognise the far greater misfortune of others.

James B. Miles, 'The Free Will Delusion'

6.1 Introduction

Any discussion of neurolaw with neuroscience students invariably veers towards the issue of culpability and determinism. In contrast, lawyers are dismissive about the issue of determinism, relegating it to the realm of philosophy and considering it irrelevant to the theory and practice of law.

Advances in neuroscience, from both human and animal studies, are changing our sense of who we are as well as the nature of life. Most drastic among these lessons is the realization that our behaviors are biologically determined and many of our behaviors can be replicated in animals. Free will advocates are caught in a bind. They accept that the mind arises entirely from brain biology and chemistry. At the same time, for free will to exist, they have to postulate a supervenient entity that exerts its control ('will') over the brain.

What neuroscience tells us is that the brain is constantly bombarded with sensory information, and it processes these inputs and adapts accordingly in a probabilistic manner. It is this intimate tango between the brain and the world that defines how we behave under every circumstance and is the essence of neuro-determinism. It is not about making intentional choices if those choices themselves are constrained by physiology and history and are unequally weighted.

In law this issue devolves rapidly into the fear of the 'my brain made me do it' excuse. How should we punish someone who is not acting of his/her own agency? The answer is simple: the same way we should treat people we deem not acting of their agency, for example, people with mental illness. We sequester them until we believe they are no longer a threat to the safety of the citizenry. This might seem a trivial deviation from current criminal law practices, but it is not. This view broadly includes all of social sciences as informed by advances in neuroscience. This might be the biggest contribution

DOI: 10.1201/9781032717791-6

of neuroscience to legal philosophy and practice. In this chapter, as a working neuroscientist, I wish to argue for the role of neuroscience in moving law from a punitive to a correctional framework (Green and Cohen, 2004). I will also argue for the need for better interactions between neuroscientists and those practicing law and more timely incorporation of discoveries in neuroscience into the theory and practice of law as well as in the mental well-being of the entire population.

6.2 Neurolaw and Culpability

The field of neurolaw is in its infancy, but its growth has been rapid in the past decade. The implication from neuroscience research that brain activity gives rise to mind and behavior, and that there is no independent agency that confers these properties, has given rise to numerous questions regarding culpability and personal responsibility.

At the core of this debate is the inescapable conundrum: there is no free will, but for every individual to exist as a productive member of a society, s/he needs to believe in its existence. None of us can function if we believe that we do not determine our destiny. This seeming paradox has given rise to a scramble to the middle from the two ends of the spectrum, between absolute determinism and a libertarian view that states that each of us is an agent capable of independent decisions for which we are responsible. From a philosophical and historical point of view, these arguments have been made exclusively in the context of morality and moral responsibility and not based on our knowledge of brain anatomy or physiology. The compatibilist tries to take the middle ground, arguing that while the brain is deterministic, the external world imposes choices upon you. However, this middle ground does not necessitate the existence of free will, just that outcomes can be varied and complex based on the choices one makes.

In criminal law, neuro-determinism is invariably equated with lack of personal responsibility and is considered a slippery slope to anarchy. Based on this, in my opinion, irrational fear, the impact of neuroscience on how we view criminality and punishment is dismissed offhand. In an article, Stephen Morse identifies a cognitive pathology [sic] called the 'Brain Overclaim Syndrome' (2006). Reminiscent of the NRA mantra, the author claims

> Brains do not commit crimes; people commit crimes. This conclusion should be self-evident, but, infected and inflamed by stunning advances in our understanding of the brain, advocates all too often make moral and legal claims that the new neuroscience does not entail and cannot sustain.

The implication is that barring dramatic and radical advances in neuroscience, it is essentially irrelevant to the theory and practice of law. However, as Green and Cohen (2004) state in their seminal article

> In modern criminal law, there has been a long tense marriage of convenience between compatibilist legal principles and libertarian moral intuitions. New neuroscience, we argue, will probably render this marriage unworkable.

A number of studies that claim evidence for (or against) neuro-determinism confuse intentionality with free will (e.g., Brass et al., 2019; Haggard, 2008; Libet et al., 1983). The human capacity for intentional decision making has never been in question, nor has our ability to make choices. A mouse can make an intentional choice of its own accord and press a lever in order to manipulate the external world (e.g., to receive a reward), but that decision, we would agree, does not define free will. The point is that as long as those choices and decisions are dependent on the underlying brain chemistry and can be modulated by altering said chemistry, arguments for the existence of free will are moot. The statement 'I can make choices' only represents free will if the entity 'I' can be separated from, or act independently of, brain physiology. If choices can be modified by simply altering brain chemistry, one cannot argue for the existence of free will unless we are willing to confer the same property to all mammals.

There is a tendency to conflate the existence of free will with the assignation of responsibility and culpability, though they aren't necessarily linked. This is because we are loath to punish those we feel do not act of their agency. This has led to the dismissal of this issue as irrelevant to the theory and practice of law.

6.2.1 Free Will – Stochastic Determinism

In the context of this chapter, one would state that based on the nature of our brain at any given time, only a specific subset of potential outcomes is possible. A stochastic process is defined here as one that has a number of random variables but with an underlying set of statistical rules that determine the probability of occurrence of a particular outcome. This is a dynamic process where each brain-world encounter alters the basal state for the next experience and hence the probability of the next outcome. Every decision we make provides us with definite options and a probability of picking one of those options (Figure 6.1). The weighted probability at any given time depends on the physiological history of the brain based on our previous experiences and consequent changes in the brain. Once we make a decision, the outcome imposes a different set of choices and different probabilities of deciding on one of them. One can see that as we go along this decision tree, we can potentially end up at vastly different places in life depending on these limited choices. While we make choices in our lives, every choice reduces the probability of making others. No amount of 'will' can turn me, a longstanding academic, into an Olympic gymnast, even if I *intend* to be one.

How brain biology affects behavior is a matter of ongoing research, and much of our knowledge comes from psychopharmacology and behavioral genetics. Illnesses like schizophrenia and depression have a strong genetic component. Twin studies demonstrate a strong genetic predisposition to schizophrenia but also show the role for other

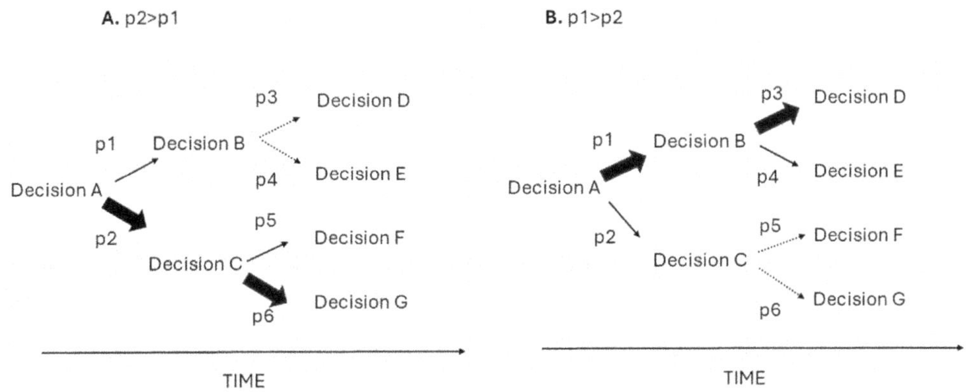

Figure 6.1 Brain decision tree.

Each decision one makes alters the likelihood of the next decision in a probabilistic manner. The weighted probability of making the next decision (p1-p6) is determined based on the brain physiology at a given time and constraints imposed by the external world. The brain state at a given time, in turn, is determined by genetics, epigenetics and prior world experiences. A. A case where p2>p1. The odds of arriving at decisions F&G downstream is relatively high while there is a low probability for making decisions D&E. B. A reverse case where p1>p2, resulting in a greater chance of downstream decisions D&E. Exogenous alterations in brain chemistry (e.g. psychiatric drugs) can alter the probability of choosing a particular course of action.

as yet unidentified factors (Hilker et al., 2018; Sullivan et al., 2003). The brain, according to neuroscience, develops as a combination of genetics and epigenetics and in response to sensory inputs from the external world. Connections are made and lost based on the Hebbian idea of activity-dependent synapse formation (Hebb, 1949). The formation of these connections is both probabilistic and driven by sensory experiences.

There are about 80 billion neurons and about 100 trillion synapses that connect them (Azeedo et al., 2009). Synaptic plasticity confers upon this organ the capacity to strengthen and weaken existing synapses based on experience, a phenomenon that underlies learning and memory. The mind arises as an intricate symphony between neuronal networks and is dependent upon the activity and regulation of underlying neurons. Genetics and epigenetics provide the set points within which behavior can be modified. Neurochemicals can alter these set points and, consequently, behavior. However, no higher-order changes can be achieved without alterations in the brain. As a corollary, changes in brain chemistry and circuits alter higher order mental processing.

In neurons, when an action potential reaches a terminal, there is a defined probability with which transmitter release can occur (less than 1). Studies indicate that this probability can vary from synapse to synapse depending on a number of factors (Branco et al., 2008). In addition, it is necessary to integrate signals from multiple synapses on a neuron before transmission can occur, and these signals have to occur within a small time window. Thus, at every point in time, there is a finite probability that an incoming signal will not be propagated. This probability is actively modulated by a number of neurotransmitters that underlie processes like attention, arousal, and learning. This probabilistic set of events is repeated across billions of synapses in order to generate behavior. Chemically modulating synaptic function, for example, with some psychiatric medications, can alter these probabilities and, consequently, behavior.

At the same time, the brain is a noisy organ. Individual neurons are also randomly spiking, and this implies that relevant signals are received over random noise, which play a role in this stochastic and dynamic process (Deco et al., 2009). Neurons form attractor networks that traverse between various stable states in a probabilistic manner. When we look at these processes over the entire brain network, it becomes apparent that brain responses are both individualistic and circumstantial and can vary at any given time. Deco et al. (2009) suggest that the stability of these brain states give rise to both normal and pathological states. These states exist as a statistical distribution, with too little stability leading to illnesses like schizophrenia and too much stability at the other end of the distribution leading to disorders like obsessive-compulsive disorders (Rolls, 2008a, 2008b).

Brain output is not a linear integration of underlying neuronal activity. At each level of complexity, the output is not a simple algebraic summation of the underlying level. While emergent properties are not always predictable from an examination of a simpler order (based on our current knowledge), it is accepted that modifying the simpler level of organization has effects on the higher levels of complexity.

At the same time, the term 'emergent properties' is also used by others to further the mind/brain dichotomy, drawing clear demarcation between the physical and mental (Moreira et al., 2018). It is flawed even to think of mind-brain interactions, as that presupposes two different entities. The brain is a dynamic system being buffeted by a number of variables in a probabilistic manner, giving rise to cognition and behavior (the mind). This does not mean that there are no rules that apply or that these rules are unknowable.

This brings us to the real world, where a stochastic brain meets a probabilistic world. Each can affect the other. For example, all of us have a finite probability that we will be hit by ongoing traffic while crossing a road. For a person with attention-deficit disorder (ADD), that probability will be higher (let's say, e.g., $p2 > p1$ in Figure 6.1). However, the same person on Ritalin (used to treat ADD, and assuming it works for the person) will have reduced that probability back to that of his/her non-ADD counterparts (e.g., $p1 > p2$ in Figure 6.1). A simple chemical modulation of the excitation-inhibition balance in the brain can alter an outcome in the external world. Similarly, people with depression often exhibit self-defeating behavior which increases the likelihood that they will fail at various tasks at work, increasing the risk of losing their jobs and furthering the behavioral cycle, while the same people might not do so on antidepressants (again assuming they are effective). In both cases, the outcomes would be the same whether the person took the drugs *intentionally* or whether forced or tricked into taking them.

The field of epigenetics shows us how external stressors (the world) can cause long-lasting changes in our DNA, programming us to become more susceptible to maladaptive behaviors like drug abuse (Hamilton and Nestler, 2019). Similarly, random events in the world that one is exposed to can increase or decrease the likelihood of repeating a behavior pattern. A random pairing of seeing someone vomiting and an ice cream parlor can result in a lifetime of aversion to the frozen dessert. The world changes the brain, and because of this change, the likelihood of that person visiting an ice cream parlor dramatically declines and, thus, changes his/her world.

It is this multiple iterative process where the brain alters experiences and is, in turn, altered by them, that shapes a personality and the likelihood of specific behaviors in response to specific circumstances. While entirely deterministic, our behavior is not a simple linear extrapolation of brain processes.

Psychiatry has had a long-standing problem with non-linear processes. Most psychiatric diagnoses as defined by the Diagnostic and Statistical Manual for Mental Disorders (DSM) are based on behavioral symptoms and not etiology. Hence, the possibility that there might be multiple means to arrive at a set of symptoms or that one brain pathway could generate multiple symptoms is not appreciated in both psychiatric diagnoses and psychotherapeutics. The inability to recognize and account for the probabilistic nature of brain processes as well as their reliance on timing is a major reason why many pharmacological agents currently in use in treatment of psychiatric illnesses are ineffective. As neuroscience research unravels mechanisms utilized in these processes, both predictability and our ability to generate better treatment paradigms are likely to improve.

Stochastic determinism states that our lives are shaped by the dynamic interaction between the brain and the external world, where both sculpt each other. Understanding the probabilistic nature of both will provide a level of predictability of human behavior that we currently do not have. While daunting at present, we are addressing similar but simpler problems with varying degrees of success. For example, scientists are getting reasonably good at predicting traffic flows, a probabilistic process that depends on the variable types of vehicles, behavior of drivers, and external conditions that might alter the flow rates. At any given instant, the status of all these variables is not known, but one can model them to arrive at practical approaches to traffic regulation (Lyapunov, 1992; Narh et al., 2016).

An individual is constantly forced to make choices. To paraphrase Sartre (1956), man is *condemned* to be free, but the choices he makes are determined by the physiological state of his brain at any given time, resulting from previous iterations and memories thereof. These processes skew the probabilities of occurrence for specific outcomes at any given instant.

6.2.2 What Does This Mean for Law?

A number of studies have highlighted problems with forensic science that lead to false convictions (Roberts, 1994; Obrien et al., 2015; Kaplan and Puracal, 2017/2018). Criminal law is often criticized for overreliance on science because it is too insecure about its capabilities. Stephen Morse criticizes belief in the power of neuroscience to drastically alter the practice of law and argues that extravagant expectations of the applicability of neuroscience to the practice of law are fueled by people averse to the concept of retributive justice (Morse, 2006, 2011). This is true to an extent. Neuroscience is not at a point where its discoveries can be used as vatic tools or even as definitive evidence of guilt or innocence. For example, it will be many years before tools like fMRI can be used as diagnostic criteria in law and even longer, if at all, before they have predictive value. However, as discussed in the following, neuroscience has a more profound and potentially transformative role in the practice of law.

Criminal law relies too much on scientists and related professionals and not so much on science. For example, this is evident in cases of mental competence and insanity defense, where the prosecution and defense might field experts who take opposing sides. The jury's decisions are then based less on science and more on soft issues like performance and believability. The problem is less with the underlying science than the inability of the legal teams to present the issues in a manner that is both accurate and accessible to jurors or to critically examine witnesses based on scientific consensus. For this, a legal team must have its own resident scientific experts and a working literacy of neuroscience to be able to cross-examine expert witnesses effectively. It is not inconceivable that in the future, the state and private law firms will have neuroscientists on their staff who can do just that.

There is a claim that criminal law is normative and based on folk psychology and not on hard science (Morse, 2011). Views on determinism based on neuroscience should, based on this claim, have very little impact on how the business of law is conducted. Here I will simply state that there is information transfer between scientific advancements and social norms. Folk psychology is not static. Our ideas about intentionality and rationality constantly change, and the new knowledge we incorporate in common parlance is just a distilled form of scientific data. Of course, it could be argued that law *must wait* for this distillation process before enacting any changes, but this misses the point. More than any other field, law must be timely in incorporating advances in science, as it deals with punishment and curbing individual freedoms, and letting people slip through statistical loopholes is not a moral option. Any state-sponsored curb on individual freedoms must meet the highest moral and scientific threshold.

6.3 The Case of Drug Addiction Laws

The concept of neuro-determinism is hard to ponder without feeling a sense of existential insecurity. After all, we would all like to believe we are captains of our ship and masters of our fate. Indeed, this need is reflected in compatibilist arguments that while previous events can determine our actions, individuals do possess autonomy, thus arguing against a completely deterministic universe.

There is a tendency to look at the role of neuroscience in a binary manner; that is, until one completely achieves a mechanistic understanding of the mind, neuroscience has nothing to offer existing legal theory or practice (Morse, 2006, 2011). Until then, it is argued that the folk-psychological model of the person drives our understanding of human behavior and, consequently, legal doctrine (Morse, 2015). This implies that legal scholarship is passive and changes in the practice of law necessarily follow changes in societal norms and practices.

Let's consider an alternative view, one that borrows the idea of *primum non nocere* from the medical field. As stated in an article by Green and Cohen (2004), 'new neuroscience will change the law, not by undermining its current assumptions, but by transforming people's moral intuitions about free will and responsibility.' This view demands a collaboration between neuroscientists and the legal profession such that the field of law plays an active part in incorporating neuroscience discoveries in a *timely* manner.

In 1954, Olds and Milner did a simple experiment (Olds and Milner, 1954). They implanted rat brains with electrodes and put them in a cage with the corners marked A, B, C, and D. Whenever an animal went to corner A, they stimulated its brain via the electrode. Most animals, subsequently, avoided corner A, except one, which seemed to prefer that corner. They discovered that in that animal, the electrode had been placed, in error, in a deeper part of the brain, in the limbic region. In these areas, it appeared that the stimulus functioned as a reward rather than punishment. They then implanted electrodes in a number of rats so as to stimulate these regions and taught the animals to deliver the shock on their own by pressing a lever. The animals stimulated their own brain more than 2000 times an hour for 24 hours without stopping. This drive for reward overrode all survival drives one assumes animals possess. When the rats were food or water deprived, the drive to press the lever to deliver electrical shocks actually increased.

Olds mapped the areas where the electrical stimulation was this strongly reinforcing and termed them 'pleasure centers.' By the early 1980s, we knew that these were the same areas that turned out to be strongly reinforcing for drugs of addiction (Goeders and Smith, 1983). Around this time, we also knew that these areas were along the dopamine pathway in the limbic region of the brain between the ventral tegmental area and the prefrontal cortex.

In parallel, Pert and Snyder (1973) published a paper demonstrating that there existed in the brain specific receptors for drugs of abuse that could mediate their actions. This study confirmed what was already proposed in 1965: that addiction is mediated by the action of drugs on brain receptors (Collier, 1965).

These two lines of research led to a consensus among neuroscientists that drugs of abuse change brain signaling and that addiction is a brain disorder. However, even as late as 1997, John Booth Davies, a British psychologist, argued that most people do drugs because they like it and have not been provided a strong enough sociological alternatives or excuses to quit. For the idea that addiction is a brain disorder to be incorporated into folk psychology, it took close to four decades, if at all. Meanwhile, the number of arrests for drug offenses rose from 50,000 in 1980 to 500,000 in 2010 (Reuter, 2013), fueled no doubt by the now-discredited war on drugs and a disregard for the emerging data from neuroscience. The question for proponents of the folk psychology argument as the only basis for law is what the acceptable number for wrongful incarceration should be in the interim period between when scientific consensus is achieved and its incorporation in the general psyche.

6.4 Moving Away from Retributivism

One tenet of retributivism is rooted in morality, accepting that those who commit wrongful acts *deserve* to be proportionately punished. The western legal system, which is followed in almost all countries today, thanks to colonialism, is based on the Judeo-Christian notion of culpability and retribution. People must suffer or be rewarded based on their evil or good deeds. Granted, we have refined the system over the last few centuries (e.g., we no longer believe it is appropriate to chop off somebody's hand for stealing), but the underlying principle has not changed. In the US that system is reinforced by current cultural norms and ideas of 'victims' rights' (an ill defined but socially popular theme) that has given rise to a mainly punitive system.

A second argument for retributive justice is that punishment acts as a deterrence both to an individual and, by example, to other potential offenders. We intuitively believe that punishment will reduce recidivism. However, data do not support this view. A recent metanalysis of 116 studies (Petrich et al., 2021) showed that custodial sanctions had no effect on, or even increased, recidivism. It is equally true that punishment of an individual does not act as a general deterrent. People who commit crimes rarely do so with the expectation of getting caught. Further, a whole host of factors like the nature of the crime, socioeconomic factors, mental states, and personality disorders undercut the strength of potential legal deterrence, resulting in a weak relationship between incarceration and crime (Ghandnoosh and Budd, 2024). On the other hand, incarceration is not benign. A recent study showed a specific increase in epigenetic markers reflective of accelerated biological aging and chronic depression suggesting that incarceration contributes to increased physical and health burden in adults (Berg et al., 2024).

A whole host of sociological factors determine an individual's odds of falling afoul of the law. Shelden and Troshynski (2020) have argued for a 'school-to-prison' pipeline driven by zero-tolerance policies and increased police presence in U.S. schools. Similarly, a case has been made for a foster care-to-prison pipeline (Yamat, 2020). There has always been demonstrated correlation between poverty levels and crime rate in the U.S. (Green, 2010; McAdams, 2010; Hagen and Peterson, 1995).

What neuro-determinism does is transfer these issues into the biological realm and, more specifically, relate them to alterations in brain structure and function. As mentioned, epigenetic changes can alter behavior (Berg et al., 2024). Histone modifications in specific brain areas upon exposure to drugs of abuse can induce lifelong behavioral abnormalities that characterize addiction (Hamilton and Nestler, 2019). A meta-analysis of 1759 studies demonstrate that early-life malnutrition can result in epigenetic modifications in cognitive and emotional areas in the brain, resulting in alterations of memory and emotional behaviors (Santos Da Silva Calado et al., 2024). Fetal alcohol exposure in humans can result in increased rates of violent behavior and crime (Roozen and Ehrhart, 2023).

Excessive stressors, via an increase in glucocorticoids, can increase epigenetic changes in brain functions that can have a large impact on behavior, especially in aggression control (Haller, 2022). Animal studies show that early life stressors (abuse, neglect, trauma) can increase alterations in emotional and social behaviors, resulting in increased risk for mood and aggressive disorders (Veenema, 2009). A more recent study demonstrates that brain region-specific alterations in methylation and binding of Sirtuin 1 (an NAD^+-dependent deacetylase) to the monoamine oxidase A promoter are associated with a sex-specific increase in early life stressor-induced aggressive behaviors (Konar et al., 2019).

The fear that relating criminal behavior to alterations in brain function would undermine the legal system and absolve a perpetrator of responsibility is unfounded. This is not a tenable argument and somewhat akin to the anti-evolution argument (if you tell people they are descended from a monkey, they will behave like one). There is some evidence that people who believe in free will have reduced tolerance for unethical behavior in others and greater propensity to be abnormally punitive (Martin et al., 2017). However, there is no credible evidence that people who do not believe in free will are prone to commit more crimes than those that do. People obey laws because, as social organisms, we are programmed to respect authority that we consider legitimate (Barton and Dunbar, 1997; Dunbar, 2009). This has arisen over generations of iterations between our brains and the external world, instilling in us concepts of right and wrong that are constantly reinforced. The sociological issues discussed previously cause early life stress and alter the brain, thus the probability that this reinforcement is weakened.

Human behavior is on a continuum. Our position on this continuum is determined by the continuous and mutual interactions of our brains with the world. Assigning an arbitrary line of knowing right from wrong or having the ability to reason fails except in the most extreme cases, for example, people with severe intellectual disability. Even patients with schizophrenia have lucid periods when they would fulfill that criterion, while 'normal' people have periods when they fail. For example, every smoker knows cigarettes can cause cancer, but for a significant fraction, that knowledge is overridden by motivational drive that is not under cognitive control. This person is, by legal standards, perfectly rational and yet acts irrationally.

Neuro-determinism tells us that each person arrives at a given time point with a distinct probability of behaving in a particular manner. The constant interaction between a stochastic brain and the world leaves each of us at a different existential point. Thus, when faced with a given situation, each person is an island unto himself or herself. The law, therefore, needs to take a holistic view incorporating sociological and psychological factors into sentencing an individual. There needs to be a collaborative structure between correctional and mental health services (Hean et al., 2017) that effectively determines whether an individual crime warrants incarceration. There is a legitimate criticism that neurotechnological advances like fMRI can provide data about a group but not about an individual. However, the same exact criticism can be leveled at the law when it confronts an individual. This obviously should rule out practices like mandatory sentencing or even a prescriptive minimum sentencing for specific crimes.

The purpose of the law is not to punish but to keep the citizenry safe. Looked at from this perspective, advances in neuroscience can have a significant impact on law. It can, for example, provide scientific bases for alternatives to the current 'one size fits all' approach to detention. There is consideration, in a number of states, to incorporate some European concepts to incarceration, especially those from the Norwegian penal system. There is a general perception in Norway that prisons are meant to train inmates to be better neighbors and that loss of freedom is punishment enough (Labutta, 2017; Crewe et al., 2023). Rather than large prison-industrial complexes, most offenders can be housed in small community centers with varying layers of security. Knowledge of behavior and the underlying brain mechanisms can definitely help in evaluating the risks of such sortings. Last, advances in neuroscience, particularly in psychopharmacology, can in some cases do away with incarceration altogether. However, this approach requires timely communication between neuroscientists and members of the legal profession.

6.4.1 Neuroscience Sans Frontieres – *A Bridge Across Silos*

Aldous Huxley (1959), in his series of lectures at Santa Barbara, stressed the need for interdisciplinary collaborators in education. He used the term *Pontifex* (used in a non-religious sense, literally meaning 'Bridge Builder'). Such people would function as a conduit across multiple disciplines. There is a need for scholars who can bridge neuroscience and law (*Pontifex iuris*, if you will) if one buys the previous argument that timely incorporation of neuroscientific concepts into law is a moral imperative.

Changes in law and society driven by neuroscience will, by necessity, be incremental, as is the nature of all scientific advances. However, as mentioned, law is morally obligated to respond to changes in a timely manner and err on the side of protecting freedoms of the citizenry. It is in this area that advances in neuroscience matter the most. As discussed, waiting for advances in the field to percolate into the general consciousness can result in large number of unfairly incarcerated people.

The key issue that determines the delay between neuroscience discoveries and their incorporation into law is communication among various disciplines. Most academic disciplines are extremely 'siloed.' Even within neuroscience, there is a communication barrier across sub-disciplines. This makes information flow extremely ineffective, taking time to percolate and increases the risk of miscommunication due to a sophisticated game of 'telephone.'

The idea of a 'Neuroscience without Borders' Center is that it brings together experts from all areas of neuroscience together with social scientists. The purpose of such an organization will be to facilitate the flow of information across disciplines. This would not be restricted to law as neuroscience impacts a number of fields like economics, psychology, philosophy, and religion. All these areas in turn impact social and cultural mores that are incorporated into folk psychology. The organization will have three mandates. First, through interactions, it will develop language that can accurately and effectively communicate concepts and paradigms across areas. This can be achieved by interactive meetings and seminars and funding collaborative projects that lead to cross-fertilization of ideas. The end goal would be to change perspectives in all directions in order to influence governmental policies based on rational neuroscience and human behavioral data. This is akin to Professor Emily Murphy's idea of collective cognitive capital that calls for the use of brain and behavior sciences–driven public policies that are aimed at the collective mental wellbeing of the population at large (Murphy, 2022; see chapter on neurolaw, this book). The second goal of such an organization would be to provide effective education in neuroscience principles and concepts for professionals outside the field. This ensures that ideas in social sciences are based on current status of scientific knowledge of brain and behavior. In the specific instance of law, it would reduce unquestioning acceptance or over-reliance on specific expert testimonies in psychiatry and neuroscience.

The third component of this organization will be aimed at changing social mindsets and attitudes by bringing neuroscience to the public. As discussed, common-sense psychology is not static; it evolves as the knowledge base of society changes. Altering public perception of what the brain does and how our behavior is affected by changes to our brain will have a profound impact on legal theories and practice.

6.5 Conclusion

If there is one lesson neuroscience teaches us, it is that human behavior is not always normative. Neither law nor the mental health profession adequately recognizes the stochastic and dynamic nature of brain–world interactions because that view is antithetical to the normative process. Currently law reconciles this dissonance by arbitrary demarcation of normal and abnormal behavior based on folk psychology. There is a common perception of rational and irrational thinking, and culpability is assigned based on that determination. Law recognizes that this approach will result in the unfair punishment of a statistical minority and accepts that as the cost of doing business. Advances in neuroscience should shift the emphasis of law from punishing an individual to keeping society safe and alter the way law does this business. However, for this approach to succeed, we need to develop a language that can be used to accurately communicate knowledge between neuroscience and other social science disciplines. A broad center that can facilitate this interaction is sorely needed.

References

Azeedo, F.A., Carvalho, L.R., Grinberg, L.T., Farfel, J.M., et al. (2009). Equal numbers of neuronal and nonneuronal cells make the human brain an isometrically scaled-up primate brain. *J. Comp. Neurol.*, 10, 532–541.

Barton, R.A., and Dunbar, R.I. (1997). Evolution of the social brain. In *Machiavellian Intelligence II: Extensions and Evaluation*, Eds. A. Whiten and R.W. Byrne, 241–308. Cambridge University Press.

Berg, M.T., Riley, K., Lei, M.-K., and Simons, R.L. (2024). Incarceration exposure, biological aging, and depression symptoms in an African American sample of older adults. *J. Aging Health*. https://doi.org/10.1177/08982643241257065

Branco, T., Staras, K., Darcy, K.J., and Goda, Y. (2008). Local dendritic activity sets release probability at hippocampal synapses. *Neuron*, 59, 475–485.

Brass, M., Furstenberg, A., and Mele, A.R. (2019). Why neuroscience does not disprove free will. *Neurosci. Biobehav. Rev.*, 102, 251–263.

Collier, H.O. (1965). A general theory of the genesis of drug dependence by induction of receptors. *Nature*, 205, 181–182.

Crewe, B., Ievins, A., Larmour, S., Laursen, J., Mjaland, K., and Schliehe, A. (2023). Nordic penal exceptionalism: A comparative, empirical analysis. *British J. Criminol.*, 63, 424–443.

Davies, J.B. (1997). *The Myth of Addiction*, 2nd edition. Harwood Academic Publishers.

Deco, G., Rolls, E.T., and Romo, R. (2009). Stochastic dynamics as a principle of brain function. *Prog. Neurobiol.*, 88, 1–16.

Dunbar, R.I. (2009). The social brain hypothesis and its implications for social evolution. *Ann. Hum. Biol.*, 36, 562–572.

Ghandnoosh, N., and Budd, K.M. (2024). *Incarceration and crime: A weak relationship*. The Sentencing Project. https://www.sentencingproject.org.

Goeders, N.E., and Smith, J.E. (1983). Cortical dopaminergic involvement in cocaine reinforcement. *Science*, 221, 773–775.

Green, J., and Cohen, J. (2004). For the law neuroscience changes everything and nothing. *Phil. Trans. R. Soc. Lond. B*, 2359, 1775–1785.

Green, S. P. (2010). Hard times, hard time: Retributive justice for unjustly disadvantaged offenders. *U. Chi. Legal F*, 2010(1), Article 4.

Hagen, J., and Peterson, R.D. (1995). Criminal inequality in America: Patterns and consequences. In *Crime and Inequality*, Eds. J. Hagen and R.D. Peterson, 19–22. Stanford University Press.

Haggard, P. (2008). Human volition: Towards a neuroscience of will. *Nat. Rev. Neurosci.*, 934–946.

Haller, J. (2022). Glococorticoids and aggression: A tripartite interaction. In *Neuroscience of Social Stress. Current Topics in Behavioral Neurosciences*, vol. 54, Eds. K.A. Miczek and R. Sinha. Springer, Cham.

Hamilton, P.J., and Nestler, E.J. (2019). Epigenetics and addiction. *Curr. Opin. Neurobiol.*, 59, 128–136.

Hean, S., Willumsen, E., and Odegard, A. (2017). Collaborative practices between correctional and mental health services in Norway: Expanding the roles and responsibility competence domain. *J. Interprof. Care*, 31, 18–27.

Hebb, D.O. (1949). *The Organization of Behaviour.* John Wiley & Sons.

Hilker, R.W., Helenius, D., Fagerlund, B., and Glenthoj, B. (2018). Heritability of schizophrenia and schizophrenia spectrum based on nationwide twin register. *Biol. Psychiatry*, 83, 492–498.

Huxley, A.L. (1959). Integrate education. In *The Human Situation-Lectures at Santa Barbara*, 1–12. Harper and Row Press.

Kaplan, A.B., and Puracal, J.C. (2017/2018). It's not a match: Why Law can't let go of junk science. *Albany Law Rev.*, 81(3), 896–939.

Konar, A., Rastogi, M., and Bhambri, A. (2019). Brain region specific methylation and Sirt1 binding changes in MAOA promoter is associated with sexual dimorphism in early life stress induced aggressive behavior. *Neurochem Int.*, 129, 1–10.

Labutta, E. (2017). The prisoner as one of us: Norwegian wisdom for Americal penal practice. *Emory Int. Law Rev.*, 31, 329–351.

Libet, B., Gleason, C.A., Wright, E.W., and Pear, D.K. (1983). Time of conscious intention to act in relation to onset of cerebral activity (readiness-potential). The unconscious initiation of a freely voluntary act. *Brain*, 106, 623–642.

Lyapunov, A.M. (1992). *The General Problem of the Stability of Motion*. Taylor & Francis, London.

Martin, N.D., Rigoni, D., and Vohs, K.D. (2017). Free will beliefs predict attitudes toward unethical behavior and criminal punishment. *Proc. Natl. Acad. Sci.*, 114, 7325–7330.

McAdams, R.H. (2010). Economic costs of inequality. *U. Chi. Legal F*, 23.

Moreira, A., Araujo, S.F., and Cloninger, R. (2018). The presentation of the mind-brain problem in leading psychiatry journals. *Braz. J. Psych.*, 40, 335–342.

Morse, S. (2006). Brain overclaim syndrome and criminal responsibility: A diagnostic note. *Ohio State J. Crim. Law*, 3, 397–412.

Morse, S. (2011). Avoiding irrational neuroLaw exuberance: A plea for neuromodesty. *Mercer L. Rev.*, 62, 837–859.

Morse, S. (2015). Neuroscience, free will, and criminal responsibility. In *Free Will and the Brain*, Ed. W. Glannon, 251–286. Cambridge University Press.

Murphy, E.R.D. (2022). Collective cognitive capital. *Wm & Mary Law Rev.*, 63, 1347–1408.

Narh, T.A., Thorpe, N., Bell, M.C., and Hill, G.A. (2016). Do new sources of traffic data make the application of Chaos Theory to traffic management a realistic possibility? *Transport Rev.*, 36, 635–658.

Obrien, P., Daeid, N.N., and Black, S. (2015). Science in court: Pitfalls, challenges and solutions. *Phil. Trans. R. Soc. B*, 370.

Olds, J. and Milner, P. (1954). Positive reinforcement produced by electrical stimulation of the septal area and othe regions of the rat brain. *J. Comp. Physiol. Psychol.*, 47, 419–427.

Pert, C.B., and Snyder, S.H. (1973). Opiate receptor: Demonstration in nervous tissue. *Science*, 179, 1011–1014.

Petrich, D.M., Pratt, T.C., Jonson, C.L., and Cullen, F.T. (2021). Custodial sanctions and reoffending: A meta-analytic review. *Crim. Justice*, 50, 353–424.

Reuter, P. (2013). Why has US drug policy changed so little over 30 years? *Crim. Justice*, 42, 75–140. The University of Chicago Press.

Roberts, P. (1994). Science in the criminal process. *Oxford J. Legal Stud.*, 14, 469–506.

Rolls, E.T. (2008a). Emotion, higher order syntactic thoughts, and consciousness. In *Frontiers of Consciousness*, Eds. L. Weiskrantz and M.K. Davies, 131–167. Oxford University Press, Oxford.

Rolls, E.T. (2008b). *Memory, Attention, and Decision-Making: A Unifying Computational Neuroscience Approach*. Oxford University Press, Oxford.

Roozen, S., and Ehrhart, F. (2023). Fetal alcohol spectrum disorders and the risk of crime. *Handb. Clin. Neurol.*, 197, 197–204.

Santos Da Silva Colado, C.M., Manhaes de Castro, R., Da Silva Souza, V., Gouvea, H.J.C.B., et al. (2024). *Nutr. Neurosci.* https://doi.org/10.1080/1028415X.2024.2361572.

Sartre, J.-P. (1956). *Being and Nothingness*. English translation by Philosophical Library Inc.

Shelden, R.G., and Troshynski, E.I. (2020). *Delinquency and Juvenile Justice in American Society*, 3rd edition, 304–305. Waveland Press, Long Grove, IL.

Sullivan, P.E., Kendler, K.S., and Neale, M.C. (2003). Schizophrenia as a complex trait – evidence from a meta-analysis of twin studies. *Arch. Gen. Psychiatry*, 60, 1187–1192.

Veenema, A.H. (2009). Early life stress, the development of aggression and neuroendocrine and neurobiological correlates: What can we learn from animal models? *Front. Neuroendocrinol.*, 30, 497–518.

Yamat, A.M. (Fall, 2020). The foster-care-to-prison pipeline. *Justice Policy J.*, 1–11.

7
Neuroscience in Psychiatry

Guido K.W. Frank and Joel Stoddard

7.1 Introduction

The field of psychiatry has undergone dramatic transformations over the past century. From a psychodynamic explanation of psychiatric disorders that was largely determined by internal conflicts and environmental factors, the field of neuroscience has helped make the underpinnings of human behavior more transparent and replace a "black box" perspective with genes, neurotransmitters and circuits that drive normal and pathological behavior (Bzdok and Ioannidis, 2019). Those changes have far-reaching implications across understanding the etiology of psychiatric disorders, how we treat those disorders and, importantly, how we teach psychiatry to the next generations of trainees.

In this chapter, we provide brief historical reviews of psychiatry and neuroscience, how neuroscience has influenced psychiatric illness models and how we may use this new understanding to develop not only new treatments but also curricula that integrate neuroscience and brain-based models of psychiatric disorders to teach students and, equally importantly, that can be used to educate patients and their families.

7.2 Historical Perspective on Psychiatry

The origin of psychiatry dates back to antiquity, societies in Africa and Europe, where mental illness was believed to be linked to gods or other supernatural phenomena and tied to spiritual rituals (Porter, 2002). Mental illness as it was perceived at that time was often seen as punishment from a higher power. This led to maltreatment of those individuals. A key problem was that the mechanisms of the mind were not accessible, which led to speculation and interpretation driven by ideology, religion or other forces. One of the first who questioned the individuals' responsibility for their aberrant behaviors were the Greek philosophers Hippocrates and Aristotle and later the Roman Celso, who proposed medical or biological models for mental illnesses and an evidence-based medicine approach (Sallam, 2010; Tondo et al., 2018). Hippocrates suggested that human disease resulted from an imbalance of the four humors: blood, phlegm, yellow bile and black bile (Smith and Smith, 2016). Furthermore, there was a distinction of madness between divine madness and pathological madness, which was considered a psychic and mental

 DOI: 10.1201/9781032717791-7

disease and foreseeable by doctors as well as curable by medications (Perdicoyianni-Paleologou, 2009).

Over the following centuries, there continued to be a distinction of unusual behaviors that were either religious in nature and driven by god or a good spirit versus others that were the results of evil spirits or a human cause, and specific rules for that distinction were established (Kemp, 2019). Importantly, during those times, the mind was still always closely linked to religion in both the Christian and Muslim world (Mora, 1980). During the Middle Ages, there was the development of so-called asylums where individuals with mental illnesses were placed (Van Renynghe de Voxvrie, 1993). The terms phrenesis, madness, melancholia, mania, epilepsy and rabies described mental disorders.

The following Renaissance period brought a shift how the human was viewed in health as well as human disease. This era that brought Humanism promoted study of the ancient cultures and a cultural and ethical rebirth, distancing itself from the Middle Ages (Grendler, 2022). During the Renaissance, doctors adapted the concept that the passions of the soul constituted disease states that could be treated by opposite passions (Van Renynghe de Voxvrie, 1993). At that time, affective disorders were considered facets of madness, whereas before that, madness was seen as a disorder of reason. In the 17th and 18th centuries, a scientific revolution occurred, spurred by advances in technology, that allowed for major advances, including understanding the human body (Osler et al., 2019). At that time, people also developed increasing compassion for the mentally ill and an interest in understanding and treating mental illness.

A truly novel era for psychiatry was entered in the 19th century with the emergence of biological psychiatry that was based on novel methods that allowed studies in neuroanatomy, neurophysiology and pharmacology (Shorter, 1997). Notable events in this century were the emergence of a framework of identifying psychological processes by Wiliam James and the identification of brain-behavioral relationships via lesion and physiology studies. At the beginning of the 20th century, Sigmund Freud developed the influential practice of psychoanalysis. This practice was based on models for the course of psychiatric illness based on mental processes that were not accessible to the conscious mind and were influenced by experiences throughout the lifespan. Early modern phenotyping of mental illness began at this time, too. This was through academic efforts such as the work of Emil Kraeplin in schizophrenia and bipolar disorder as well as by societal demand for a coherent classification system for mental disorders.

In parallel, psychologists developed methods for the operationalization and study of behavior as well as formal methods for measuring psychological constructs, that is, psychometrics. The persistent tension between behaviorists and psychoanalysts on how to identify latent mental states culminated in the cognitive revolution of the 1950s, which led to modern methods for inferring and measuring mental operations. Cognitive psychology and neuroscience were formally integrated in the 1970s into the field of cognitive neuroscience, providing a foundation for the current approach to psychiatric neuroscience.

Also in the 1970s, calls for a structured, reliable diagnostic approach for the definition of mental disorders grew to intense pressure, culminating in the 1980 publication of the *Diagnostic and Statistical Manual for Mental Disorders, 3rd edition* (DSM). Primarily, disorders were defined by clinical judgement of a set of symptoms and functioning, without a determination of cause or biologic examination, which was not feasible at the time. The DSM had a major influence on our current definitions of mental disorders through its clinical and legal use as well as its translation to formal interviews for classifying individuals in clinical trials and neuroscientific studies.

The end of the 20th century brought rapid advancement with the development of neurotransmitter-based hypothesis for mental disorders, the development of human functional brain imaging that allowed in vivo study of brain function and the genetic revolution with identification of the human genome (Kandel, 2013). In addition, there was an increase in interventions, both biological and behavioral, as well as methods for assessing their efficacy and mechanism of action, allowing us to establish their usefulness and 'active ingredients.' Through this process, neural hypotheses could be generated.

After decades of use, the limitations of the DSM system have become clear. It was rare for investigators to discover the etiology of a disorder, such as Rett's disorder, and so the biological basis for most DSM-based disorders has not been established. Given the apparent heterogeneity and their potential lack of validity with respect to biology, a novel approach via large-scale data collection and classification was suggested (Cuthbert and Insel, 2013). A framework for doing so, the Research Domain Criteria (RDoC), was proposed by the National Institute of Mental Health to facilitate the definition of phenomena related to mental health and based on their biology.

Facilitating this work has been the focus on the acquisition of large amounts of data and the rapid development of computational power and algorithms. These will likely yield additional insights into the pathophysiology of mental illness and individualized predictors of health. The use of machine learning helps to identify patterns that may be purely data driven (Wang et al., 2020). In addition, the development of generative artificial intelligence promises to improve personalized medicine by integrating contextual factors for the care of patients (King et al., 2023). Finally, both simple algorithms and complex system models have been developed to describe theoretical processes. These include biologically realistic models of brain function, developed to test specific hypotheses about system neural function as well as to simulate 'what-if' scenarios.

These recent advances, and the limitations of our current diagnostic system, have facilitated the modern approach, which has several elements. The reliance on the current diagnostic system is increasingly augmented with precision medicine approaches. Precision psychiatry refers to a patient-oriented diagnostic and treatment approach considering validated individualized factors, such as known genetic or biological variants or personalized risk scores based on multiple features outside of diagnostic classification. Treatment development continues to take a mechanistic approach; the process helps identify cognitive, neural, social and other mechanisms that contribute to mental illness. The recent use of rapid-acting antidepressants has spurred numerous new hypotheses of mechanism of action. Alternatively, treatments have also been developed considering the mechanism of illness, such as modern neuromodulation. (Another important growing aspect of a treatment development focus ensures community engagement, participation and acceptance.) To guide mental health professionals, open-source educational content has become widely available, such as the National Neuroscience Curriculum Initiative (https://nncionline.org/), which is an excellent resource for further exploration of cutting-edge neuroscience-based formulations of mental health issues.

7.3 Historical Perspective on Neuroscience

While psychiatry is focused on mental disorders and their underlying neurobiology, neuroscience focuses on the study of the structure and function of the nervous system in health and in disease. There are many parallels between the development of neuroscience and psychiatry from antiquity to modern age, since they overlap in their focus on the nervous system and often depend on each other to identify and elucidate neural processes, such as cognition or behavior, of interest. Some of the earliest studies of brain function focused on brain injuries and how they affected a person's behavior. Hippocrates already hypothesized that the brain was the organ that controls the body (Finger, 2000). Similarly, and not surprisingly, the Renaissance was also a time of great progress for the emerging field of neuroscience. Increasingly anatomical dissections of the body and brain took place to develop maps of the body's organ systems. This was paralleled by the invention of the microscope and starting to study the human body on the cellular level (Finger, 2000). This led to a cellular- and neuron-based understanding of the human nervous system developed by Cajal and Golgi, using staining techniques of individual cells and developing models for neurocircuitry (Glickstein, 2006). Neuroscience rapidly developed further with novel techniques that entered the field with electron microscopy, neurochemical assays, genetic analyses and optogenetics, which allow in vivo cell manipulation and the study of neural circuits (Ganesana et al., 2017).

7.4 Toward Brain Circuit–Based Human Behaviors

After a very rudimentary beginning, the way we understand human behavior has undergone dramatic changes over the past century. We are now at the point where research has indicated specific circuits involved in our thoughts, experiences, and actions. These are the core clinical features of interest to psychiatrists. Moreover, clinical neuroscience has advanced enough to bring about the development of brain-based interventions for mental disorders, such as deep brain stimulation for obsessive compulsive disorder or transcranial magnetic stimulation for depressive disorders.

The brain is composed of billions of cells that form anatomically organized units with defined functions that connect with other brain regions via neurons. Those connections, via electric and neurochemical transmission, govern our emotions, cognitions and actions. While neuroscience has identified those circuits in animal and in vitro studies, they can be now also be studied in noninvasive ways in humans using techniques such as functional magnetic resonance imaging (fMRI), electroencephalography (EEG) and positron emission tomography (PET). Those methods that have been developed and refined over the past half century are key to mapping circuits in healthy individuals as well as psychiatric populations (Sporns, 2022). These techniques identify large regions or neural signaling patterns. Simultaneously, techniques have been developed to examine more closely the cellular and molecular level in animal studies. For example, the introduction of optogenetics has helped further refine models of brain function, with the ability to distinctly target specific neurons and genes using viral vectors and elicit neuronal activation. These techniques give us a better understanding of how specific circuits act.

In addition to structural and functional connections, advances in understanding communication in the central nervous system underlie much of our drug development. There are several major neurotransmitter systems that underlie chemical signal transduction in the brain. Those include amino acid neurotransmitters, acetylcholine, catecholamines, serotonin and histamine, as well as neuropeptides that act on their respective neurotransmitter receptors (Cooper et al., 2003). Those circuits may be organized along broad tracts with neurons originating in the brain stem, such as the raphe nuclei for the serotonin system or the midbrain and substantia nigra for dopaminergic circuits, and extending from there throughout the brain. Other neurotransmitters such as gamma-aminobutyric acid (GABA) and glutamate are ubiquitous but without topical distribution. Most neurotransmitters have characteristic functions that broadly help explain aspects of cognition, emotion, and behavior.

However, despite those advances, the connection from cells to neurons and circuits driving human behaviors remain complex and incompletely understood. Research has mostly been done in animals and is not always directly translational to humans, and while a study can identify certain pathways involved in behavior, the many potential confounding factors such as acute environmental influences are difficult to model in those study designs.

Several specific neural circuits that are critical to human behavior and psychiatric disorders have been identified. For instance, there is corticostriatal circuitry that has a central role in motivation and reward processing. This circuitry connects the prefrontal cortex, the ventral striatal nucleus accumbens and the ventral tegmental area in the midbrain (Haber, 2016).

Afferent projections from the prefrontal cortex connect to the striatum, while the mesocortical and mesolimbic pathways from the midbrain provide dopaminergic input to the frontal cortex and ventral striatum, respectively. Those circuits are important in processing expectation and outcome contingencies, the so-called prediction error model, via phasic and tonic dopamine neuronal response (Haber and Knutson, 2010). In general, activation of the corticostriatal circuit has been associated with reward-seeking behaviors such as eating, social interaction and substance use. Over the past decade, a more refined model has emerged that does not simply link reward seeking with those brain circuits. Context-dependent learning and individual cognitive biases are key to motivate behavior and associated dopaminergic response. For instance, food approach is triggered when feeling safe via rostral nucleus accumbens dopamine D1 receptor activation. Feeling unsafe, on the contrary, triggers dread and avoidance toward food via the combined activation of dopamine D1 and D2 receptors in the caudal nucleus accubens (Richard and Berridge, 2013).

Another important circuit for psychiatric psychopathology and pathophysiology is the corticolimbic circuit, which has a central role in emotion regulation. This circuit includes the amygdala, hippocampus and parts of the prefrontal cortex, including the anterior cingulate. The amygdala is an important region for vigilance and threat detection and responds to fear- and anxiety-inducing stimuli (LeDoux, 2003). The hippocampus integrates context with emotional response (Chang and Liang, 2017), and the frontal cortical regions use the input from the amygdala and hippocampus to form a flexible behavioral response based on the individual circumstance and the organism's need at the moment (Rusbridge, 2020). While the amygdala responds to threat and triggers a fight-or-flight response, prefrontal afferents to the amygdala are meant to evaluate the perceived threat and, if needed, control or moderate amygdala activation (Quirk et al., 2003).

Both of those circuits develop at different time scales through childhood and adolescence, which has implications for behavior (Cavanna, 2013). The striatal reward circuitry develops early in life, driving goal-directed behavior in childhood and adolescence that is less controlled by, for instance, fear of consequences. The latter is processed by the prefrontal cortex, which executes top-down control but matures later in adolescence. This imbalance is considered responsible for impulsive behaviors in youth.

In summary, there are distinct neurocircuits that have been identified in animal and human research that reliably get activated when the individual engages certain cognitive or emotional processes. Tasks that trigger those circuits can be used in psychiatric research to identify vulnerabilities, inefficiencies or deficits in information and emotion processing across psychiatric disorders and to evaluate neurotransmitter systems that are part of their pathophysiology.

7.5 Neuroscience and Psychiatric Illness Models

There has been an increasing understanding of how neurocircuitry drives psychiatric illnesses. This has offered many new insights into the complex interplay between brain function and mental health. Psychiatric disorders, such as depression, schizophrenia, bipolar disorder and anxiety disorders, are characterized by disruptions in mood, cognition, perception and behavior, profoundly impacting the lives of those affected. While the precise etiology of these disorders remains multifaceted and multifactorial, research has increasingly focused on unraveling the neurobiological underpinnings that contribute to their onset and progression.

With the identification of the major neurotransmitter systems involved in brain neurotransmission, psychiatry entered a new era of understanding psychopathology and also developing novel treatments. Research has shown that neurotransmitters such as serotonin, dopamine, norepinephrine and gamma-aminobutyric acid play key roles in psychiatric illnesses by regulating mood, anxiety, stress response, value computation and motivation (Nestler and Hyman, 2010). Altered response or neuronal function of these systems has been associated with, for instance, unipolar or bipolar mood disorders or anxiety disorders, where altered serotonin tone or release was hypothesized to contribute to dysfunctional behavior. Dopamine transmission, on the other hand, was associated with schizophrenia, substance use disorders or attention deficit hyperactivity disorder (Grace, 2016; Belujon and Grace, 2017). However, those associations were in the beginning somewhat simplistic, and there is now an emerging picture that there is neurotransmitter cross-talk as opposed to a one neurotransmitter–one disease model.

The advancements in neuroimaging such as structural and functional MRI, magnetic resonance spectroscopy (MRS), PET and advanced EEG have helped identify structural and functional changes associated with psychiatric illnesses. Those studies have revealed regional or pervasive changes in brain structure and volume, white matter tract connectivity or regional alterations in brain activation in response to tasks that test specific domains of behavior.

For instance, earlier studies found changes in the volume and function in the prefrontal cortex, amygdala and hippocampus have been associated with depressive or bipolar disorders (Drevets, 2001) or altered dopaminergic pathways and cortical-subcortical circuits in schizophrenia that were related to psychotic symptoms and cognitive deficits (Howes and Kapur, 2009; Anticevic et al., 2015). Since then, there has been an explosion of neuroscience-based research in psychiatry. A literature search only using the terms psychiatry and neuroscience in PubMed resulted in over 115,000 hits at the time of writing this chapter. Reviews on neuroscience-informed studies in psychiatry have led to many novel results and a much better understanding of psychiatric illnesses. For instance, today one might visit neurosynth.org for an on-the-fly metanalysis of functional neuroimaging studies for a specific condition or behavior. Another effort to synthesize this literature with respect to mental health is the aforementioned RDoC, where cognitions, emotions and actions are organized by neural systems.

Current trends in our understanding of mental disorders and their treatment development increasingly involve an understanding of their neurobiology. Neurobiological factors may underlie major depressive disorder, but they are diverse and often nonspecific, indicating that the disorder has diverse origins and likely typologies that need further understanding (Singh et al., 2024; Cui et al., 2024). Still, common elements of disorders can be found. For example, the bipolar disorder spectrum has been consistently associated with altered reward sensitivity and goal value processing and various cortical and subcortical structural and functional alterations (Phillips, 2023). Genetics research has strongly implicated heritability in attention deficit hyperactivity disorder and brain circuits involved in dopamine neurotransmission, which was supported by medications that elevate catecholamine tone in the brain alleviating symptoms (da Silva et al., 2023). Other neuroscience-based translational research has studied fear processing to develop so-called brain biotypes and identify biomarkers of fear, anxiety or trauma-related disorders (McClellan France and Jovanovic, 2023). Often, advances in neuroscience will also lead us to a new understanding of mental disorders. One emerging field has been the study of the interface between the nervous system and our immune response. Neuroimmunological studies developed a framework of how inflammation and immune dysregulation during development could contribute to low mood and altered reward processing in depressive disorders (Nusslock et al., 2024). More concretely, clinical guidelines have been emerging for the diagnosis and treatment of autoimmune neurological syndromes (Abboud et al., 2021).

Another important observation has been that the median age of onset of psychiatric disorders is relatively early in life at 11 years of age for anxiety and impulse-control disorders but later at age 20 years for substance use and age 30 years for mood disorders (Kessler et al., 2005). With half of all lifetime psychiatric disorders starting by age 14 and three fourths by age 24, it is likely that developmental factors and maybe the interaction between biological traits and environmental stressors have an important role in developing a psychiatric disorder.

All in all, there has been much work done to use neuroscience-informed methods and translational approaches to research psychiatric disorders. NIMH's RDoC project includes a matrix that integrates domains of behavior, such as positive and negative valence, sensorimotor, social and cognitive processes, and studies those across diagnoses across the development and lifespan based on underlying genes, molecules and cells, brain circuits and physiology (Cuthbert, 2014; Insel, 2014). This has led to many new discoveries and models to help explain their underpinnings. The stance that mental disorders are brain based, and not a moral failing, has helped reduce stigma of mental illness and create medical models that can be used in psychoeducation. On the flip side, the direct translation of neuroscience findings into effective new treatments has been difficult. While some novel medications or treatments such as ketamine for major depression or OCD or neurostimulation interventions have been developed, there is still a lack of effective treatments to tackle the often-discussed mental health crisis (Gunnell et al., 2018). This raises the question of whether we have always been putting research efforts toward the most valuable targets.

7.6 Neuroscience in Examples of Psychiatric Disorders

7.6.1 The Neuroscience of Aggression

Broadly, psychiatric syndromes as defined by the *Diagnostic and Statistical Manual for Mental Disorders* or International Classification of Diseases (ICD) are complex, heterogenous phenotypes that are largely socially constructed. Due to the lack of our understanding of their underlying mechanisms, they are termed disorders as opposed to diseases or syndromes (DSM-5) and largely affect the social, affective, cognitive and behavioral domains. As such, very few disorders have a clearly defined neural basis. On the other hand, specific symptoms or signs, such as aggression, may often be tractable for study. Indeed, treatments more often target a problematic aspect of a disorder, such as hallucinations in schizophrenia, rather than all aspects of the disorder.

Though aggression is a complex behavior, defined as a behavior intended to hurt another, it may be reasonably well studied. It is a reliably measured phenotype that exhibits stability through life (Olweus, 1979). It is also observable and measurable across cultures and species. Finally, it may be reliably elicited through experimental manipulations, such as serotonin deprivation, hypothalamic stimulation and frustration.

Broad categories of aggression have been observed, with two of the most commonly considered being reactive and proactive. Reactive aggression is a stimulus response. For example, it may be a response to threat along the classic 'freeze–flight–fight' reaction sequence. Proactive aggression is part of an overall plan used to achieve a goal. For example, proactive aggression may be used in predation. While these categories are distinguishable psychometrically, they do have blurry boundaries.

Given the complexity of aggression, many researchers limit their inferences to specific stimulus–behavior associations which may be experimentally verified. One of the most well-developed hypotheses in aggression is the frustration-aggression hypothesis (Berkowitz, 1989), which began with the observation that individuals who tend to be aggressive also tend to be more sensitive to frustration. Frustration is a state induced by the omission of an expected reward or delivery of an unexpected punishment (Berkowitz, 1989).

A restricted construct of frustration is frustrative non-reward, which is amenable to experimental systems in humans and animals (Amsel and Ward, 1954). Frustrative non-reward is the frustration response specifically to reward omission. It has been included in the NIMH RDoC due to its relevance to psychopathology. Because it is clearly defined by a stimulus, well-operationalized experiments in humans and animals can induce an expectation and then violate it. Mice undergoing frustrative non-reward paradigms exhibit aggression-related behaviors. Recent experiments using c-Fos imaging to examine its neural correlates demonstrate widespread changes in neural activity after frustrative non-reward. These include limbic and basal ganglia regions involved in behavioral and physiologic activation that have been previously implicated in human studies and animal studies. At this point, such models will allow for distinct experiments manipulating specific neural pathways and their association with aggressive responses.

For humans, frustrative non-reward may be induced by a variety of tasks often involving monetary incentives. For ethical reasons, aggressive behaviors are limited to proxies, that is, surrogate indicators such as physiologic arousal and motor response, or allowing aggression with minimal harm, such as point subtraction in a game. In such experiments, pathways of neural activation are like mice in that they involve affect generation, motor generation and associative and regulatory areas. The requirement in those experiments is that expectation of reward clearly implicates areas that encode value, with this value construct representing the magnitude of expectation of a desired reward. In humans, value encoding is highly abstracted and depends on the orbital frontal cortex (OFC). This may lead to complex phenotypes and blurring of boundaries between neatly separable constructs. For example, one influential theory posits that individuals with callous unemotional traits, or psychopathy, are often insensitive to the distress of others, which may be related to low activation of the OFC to distress cues. However, the same theory suggests that their aberrant encoding of value in general leaves them more prone to experience frustration and thereby explains the greater degree of reactive (hot-blooded) and not just proactive (cold-blooded) aggression observed in these individuals relative to the general population.

While frustrative non-reward is reasonably narrow in terms of is operationalization, no discussion of the neurobiology of aggression would be complete without mentioning classic investigations and the importance of this case in the intersection between society and neuroscience. Classic investigations identified aberrant neural activity or lesions in individuals with a tendency to aggress. This includes the famous case study of the accident that damaged the orbitofrontal and medial prefrontal cortices of Phineas Gage (Damasio et al., 1994). After the accident, Mr. Gage became more aggressive. Studies by Adrian Raine in the 70s and 80s demonstrated decreased metabolism in prefrontal areas in violent offenders (Raine et al., 1998). Of course, most people can appreciate that aggressive behaviors are elevated in some at the extreme ages of life, corresponding to periods of relatively lower functioning of the phylogenetically younger regulatory cortex.

Taken to an extreme, any biological pathways identified to be associated with or mediate aggression are taken by proponents of biologic determinism, such as Cesare Lombroso, to influence policy or law. Other extremes include sociological and psychological determinism. While the brain is the basis of complex behavior, it is plastic to these domains of influence. As noted, very few disorders have any identifiable, unique biological basis. Moreover, discrete biological lesions at any level, from genes to behavior, have a probabilistic expression of various signs and symptoms. Consider the uncertain behavioral outcomes of individuals with a relatively discrete MECP2 lesion underlying the pleiomorphic features of Rett syndrome or chromosome 22q11.2 microdeletion leading to DiGeorge syndrome and potentially a schizophrenia phenotype. Even more distal lesions, like stroke in anatomically analogous areas in individuals, may have uncertain behavioral and psychological outcomes. Mental health practitioners seek to identify contributors to functional impairment in life domains or distress. However, the identification of biological pathways will continue to help us improve diagnosis, treatment and prognosis.

7.6.2 The Neuroscience of Eating Disorders

Eating disorders (EDs) are severe psychiatric disorders with high mortality (Crow et al., 2009). Several distinct disorders have been identified, including anorexia nervosa (AN), characterized by severe underweight; bulimia nervosa (BN), presenting with episodic binge eating and purging; and binge eating disorder (BED), with episodes of overeating without compensatory behaviors (American Psychiatric Association, 2022). Those disorders are cross-transmitted in families, indicating shared underlying neurobiology (Robinson et al., 2020).

Animal models for EDs implicated the monoamine neurotransmitters dopamine and serotonin in food restriction or excessive intake. The activity-based anorexia (ABA) model associated excessive dopaminergic activation and overexpression of the dopamine D2 receptor with weight loss (Welch et al., 2021; Beeler et al., 2021). Environmental factors triggering stress response alter dopamine and serotonin pathways; those alterations may be enduring and perpetuate altered eating (Carr, 2002, 2011; Beeler and Burghardt, 2021; Reed et al., 2021; Giunti et al., 2023). Animal models for binge eating behaviors revealed acute elevations in dopamine neurotransmission but a decrease in striatal dopamine receptor distribution after prolonged binge eating coupled with anxiety when deprived of binge eating (Johnson and Kenny, 2010; Avena et al., 2012). Human studies demonstrated elevated serotonin 1A-receptor binding in AN and BN regardless of ill or recovered state; in contrast, serotonin 2A receptors appeared to shift from normal to low expression during recovery, suggesting dynamic adaptations. Additionally, dopamine D2 receptor availability was normal in ill AN and BN but elevated in the striatum in AN after recovery (Frank et al., 2005; Broft et al., 2015).

While those studies indicated neurotransmitter involvement in EDs, linking them to behavior or functional outcome has been difficult. Functional magnetic resonance brain imaging using tasks that involve dopaminergic neurotransmission indicated several important findings. First, dopamine-associated brain circuit response is associated in opposite directions with low- and high-weight eating disorders, with hyper-responsiveness in under- and hypo-responsiveness in overweight individuals (Frank et al., 2021). Second, anxious traits and fear conditioning have an important role in EDs and have been associated with elevated amygdala activation (Frank et al., 2023). Experimental conditioned fear to high caloric taste stimuli also predicted long-term weight outcomes in AN (Gorrell et al., 2024). Animal research indicated that food approach and avoidance depend on conditioned behavior and are mediated via the rostral or caudal nucleus accumbens, respectively, and dopamine D1 or the joint activation of D1 and D2 receptors (Richard and Berridge, 2011). Recent human brain imaging from our group supports this model in AN when trying to avoid eating high caloric food (Weider et al., 2024). While effective neurobiological treatments for EDs are still scarce, those models help educate patients and families to better understand those disorders and develop treatment targets for further study (Kirchberg et al., 2024).

7.7 Neuroscience in Psychiatry Training

The burgeoning amount of neuroscience-based research but its limitations with respect to development of novel treatments raises the question of whether or to what extent psychiatry training should include training in neurobiology, neurocircuits and their relationships with mental health or illness. In fact, some have questioned to what extent neuroscience can be used or be useful in a better understanding of psychiatry and considered it reductionist (Rose, 2016). Many trainees have traditionally focused on interview, assessment and diagnosis, psychotherapy and how to use medication to alleviate symptoms. For many, additional training in neuroscience appears burdensome, and they do not recognize the need to learn this aspect of psychiatry and integrate in their patient interactions. This, however, is reductionistic and ill guided in our view. A neuroscience-based understanding allows one to present the patient with a truly holistic view of their illness that can integrate biological traits, environmental influences and neurobiological contributions to the illness. This helps reduce stigma and take psychiatry from a lack of understanding to a transparent subspecialty of medicine that is organ based and can be treated using evidence-based interventions. Neuroscience and development-based understanding of psychiatric illness also allow to better use psychopharmacology. For instance, neurotransmitter systems change with age and body weight, and therefore medications may need to be adjusted. In general, to apply the bio-psycho-social model for psychiatric disorders, one needs to be aware of the biological aspects, just as one assesses trauma or other factors. It is therefore imperative that neuroscience be seen as an integral part of psychiatric research and treatment that must be important for clinician training. Patients are also able to learn much from the internet, and a savvy patient might have more neurobiological knowledge than a psychiatrist. If that is the case, the patient might not trust the provider who may have less knowledge. Organizations such as the American Psychiatric Association (APA) or American College of Neuropsychopharmacology (ACNP) have made strides to enhance neuroscience training of clinicians, and residency training programs now regularly include lecture series on the neuroscience behind or underlying mental illness. However, this training needs to be adjusted constantly to make it worthwhile and useful for the trainee and clinician to use at the bedside (Cooper and Walker, 2021). A major problem that we see is that the vastness of new information becomes more and more difficult to review and integrate, with studies frequently contradicting each other, while trainees need clearly to understand information that can be relatively easily translated into patient-oriented care.

7.8 Conclusions

The field of psychiatry has undergone dramatic changes from its origins to today by incorporating psychodynamic theory, psychological and behavior models and more recently pathophysiology based on biology-driven brain circuits and neurotransmitters that the field of neuroscience has identified. The next step in this development will be to further develop and refine those models and work toward a holistic gestalt that represents individual diagnoses and syndromes while bridging between those based on behavior and neurobiology. Most importantly, this work must be directed toward a perspective where a person is viewed as an individual with specific needs to support patient-centered recovery and health.

References

Abboud, H., Probasco, J.C., Irani, S., Ances, B., Benavides, D.R., Bradshaw, M., Christo, P.P., Dale, R.C., Fernandez-Fournier, M., Flanagan, E.P., Gadoth, A., George, P., Grebenciucova, E., Jammoul, A., Lee, S.T., Li, Y., Matiello, M., Morse, A.M., Rae-Grant, A., Rojas, G., Rossman, I., Schmitt, S., Venkatesan, A., Vernino, S., Pittock, S.J., Titulaer, M.J., Autoimmune Encephalitis Alliance Clinicians, N., 2021. Autoimmune encephalitis: proposed best practice recommendations for diagnosis and acute management. *Journal of Neurology, Neurosurgery and Psychiatry* 92(7), 757–768. PMC8223680. Access Number: 33649022.

American Psychiatric Association (2022). *Desk reference to the diagnostic criteria from DSM-5-TR*. Washington, DC, American Psychiatric Association Publishing: 1, online resource.

Amsel, A., Ward, J.S., 1954. Motivational properties of frustration: II. Frustration drive stimulus and frustration reduction in selective learning. *Journal of Experimental Psychology* 48(1), 37–47. Access Number: 13174731.

Anticevic, A., Savic, A., Repovs, G., Yang, G., McKay, D.R., Sprooten, E., Knowles, E.E., Krystal, J.H., Pearlson, G.D., Glahn, D.C., 2015. Ventral anterior cingulate connectivity distinguished nonpsychotic bipolar illness from psychotic bipolar disorder and schizophrenia. *Schizophrenia Bulletin* 41(1), 133–143. PMC4266289. Access Number: 24782562.

Avena, N.M., Bocarsly, M.E., Hoebel, B.G., 2012. Animal models of sugar and fat bingeing: relationship to food addiction and increased body weight. *Methods in Molecular Biology* 829, 351–365. Access Number: 22231826.

Beeler, J.A., Burghardt, N.S., 2021. The rise and fall of dopamine: a two-stage model of the development and entrenchment of anorexia nervosa. *Frontiers in Psychiatry* 12, 799548. PMC8787068. Access Number: 35087433.

Beeler, J.A., Mourra, D., Zanca, R.M., Kalmbach, A., Gellman, C., Klein, B.Y., Ravenelle, R., Serrano, P., Moore, H., Rayport, S., Mingote, S., Burghardt, N.S., 2021. Vulnerable and resilient phenotypes in a mouse model of anorexia nervosa. *Biological Psychiatry* 90(12), 829–842. PMC7855473. Access Number: 32950210.

Belujon, P., Grace, A.A., 2017. Dopamine system dysregulation in major depressive disorders. *International Journal of Neuropsychopharmacology* 20(12), 1036–1046. PMC5716179. Access Number: 29106542.

Berkowitz, L., 1989. Frustration-aggression hypothesis: examination and reformulation. *Psychological Bulletin* 106(1), 59–73. Access Number: 2667009.

Broft, A., Slifstein, M., Osborne, J., Kothari, P., Morim, S., Shingleton, R., Kenney, L., Vallabhajosula, S., Attia, E., Martinez, D., Timothy Walsh, B., 2015. Striatal dopamine type 2 receptor availability in anorexia nervosa. *Psychiatry Research* 233(3), 380–387. PMC5055757. Access Number: 26272038.

Bzdok, D., Ioannidis, J.P.A., 2019. Exploration, inference, and prediction in neuroscience and biomedicine. *Trends in Neurosciences* 42(4), 251–262. Access Number: 30808574.

Carr, K.D., 2002. Augmentation of drug reward by chronic food restriction: behavioral evidence and underlying mechanisms. *Physiology and Behavior* 76(3), 353–364. Access Number: 12117572.

Carr, K.D., 2011. Food scarcity, neuroadaptations, and the pathogenic potential of dieting in an unnatural ecology: binge eating and drug abuse. *Physiology and Behavior* 104(1), 162–167. PMC3107914. Access Number: 21530562.

Cavanna, A.E. (2013). *Frontal lobe: anatomy, functions and injuries*. New York, Nova Science.

Chang, S.D., Liang, K.C., 2017. The hippocampus integrates context and shock into a configural memory in contextual fear conditioning. *Hippocampus* 27(2), 145–155. Access Number: 27806432.

Cooper, J.J., Walker, A.E., 2021. Neuroscience education: making it relevant to psychiatric training. *Psychiatric Clinics of North America* 44(2), 295–307. Access Number: 34049650.

Cooper, J.R., Bloom, F.E., Roth, R.H., 2003. *The biochemical basis of neuropharmacology.* Oxford, UK, Oxford University Press.

Crow, S.J., Peterson, C.B., Swanson, S.A., Raymond, N.C., Specker, S., Eckert, E.D., Mitchell, J.E., 2009. Increased mortality in bulimia nervosa and other eating disorders. *American Journal of Psychiatry* 166(12), 1342–1346. Access Number: 19833789.

Cui, L., Li, S., Wang, S., Wu, X., Liu, Y., Yu, W., Wang, Y., Tang, Y., Xia, M., Li, B., 2024. Major depressive disorder: hypothesis, mechanism, prevention and treatment. *Signal Transduction and Targeted Therapy* 9(1), 30. PMC10853571. Access Number: 38331979.

Cuthbert, B.N., 2014. Translating intermediate phenotypes to psychopathology: the NIMH Research Domain Criteria. *Psychophysiology* 51(12), 1205–1206. Access Number: 25387702.

Cuthbert, B.N., Insel, T.R., 2013. Toward the future of psychiatric diagnosis: the seven pillars of RDoC. *BMC Medicine* 11, 126. PMC3653747. Access Number: 23672542.

Damasio, H., Grabowski, T., Frank, R., Galaburda, A.M., Damasio, A.R., 1994. The return of Phineas Gage: clues about the brain from the skull of a famous patient. *Science* 264(5162), 1102–1105.

da Silva, B.S., Grevet, E.H., Silva, L.C.F., Ramos, J.K.N., Rovaris, D.L., Bau, C.H.D., 2023. An overview on neurobiology and therapeutics of attention-deficit/hyperactivity disorder. *Discover Mental Health* 3(1), 2. PMC10501041. Access Number: 37861876.

Drevets, W.C., 2001. Neuroimaging and neuropathological studies of depression: implications for the cognitive-emotional features of mood disorders. *Current Opinion in Neurobiology* 11(2), 240–249. Access Number: 11301246.

Finger, S., 2000. *Minds behind the brain: a history of the pioneers and their discoveries.* Oxford; New York, Oxford University Press.

Frank, G.K., Bailer, U.F., Henry, S.E., Drevets, W., Meltzer, C.C., Price, J.C., Mathis, C.A., Wagner, A., Hoge, J., Ziolko, S., Barbarich-Marsteller, N., Weissfeld, L., Kaye, W.H., 2005. Increased dopamine D2/D3 receptor binding after recovery from anorexia nervosa measured by positron emission tomography and [11c]raclopride. *Biological Psychiatry* 58(11), 908–912. Access Number: 15992780.

Frank, G.K.W., Shott, M.E., Pryor, T., Swindle, S., Nguyen, T., Stoddard, J., 2023. Trait anxiety is associated with amygdala expectation and caloric taste receipt response across eating disorders. *Neuropsychopharmacology* 48(2), 380–390. PMC9750993. Access Number: 36100656.

Frank, G.K.W., Shott, M.E., Stoddard, J., Swindle, S., Pryor, T.L., 2021. Association of brain reward response with body mass index and ventral striatal-hypothalamic circuitry among young women with eating disorders. *JAMA Psychiatry* 78(10), 1123–1133. PMC8246338. Access Number: 34190963.

Ganesana, M., Lee, S.T., Wang, Y., Venton, B.J., 2017. Analytical techniques in neuroscience: recent advances in imaging, separation, and electrochemical methods. *Analytical Chemistry* 89(1), 314–341. PMC5260807. Access Number: 28105819.

Giunti, E., Collu, R., Dedoni, S., Castelli, M.P., Fratta, W., Scherma, M., Fadda, P., 2023. Food restriction and hyperactivity induce changes in corticolimbic brain dopamine and serotonin levels in female rats. *Behavioural Brain Research* 444, 114374. Access Number: 36863461.

Glickstein, M., 2006. Golgi and Cajal: the neuron doctrine and the 100th anniversary of the 1906 Nobel Prize. *Current Biology* 16(5), R147–R151. Access Number: 16527727.

Gorrell, S., Shott, M.E., Pryor, T., Frank, G.K.W., 2024. Neural response to expecting a caloric sweet taste stimulus predicts body mass index longitudinally among young adult women with anorexia nervosa. *Biological Psychiatry: Cognitive Neuroscience and Neuroimaging* 9(3), 298–304. PMC10811282. Access Number: 37506848.

Grace, A.A., 2016. Dysregulation of the dopamine system in the pathophysiology of schizophrenia and depression. *Nature Reviews: Neuroscience* 17(8), 524–532. PMC5166560. Access Number: 27256556.

Grendler, P.F. (2022). *Humanism, universities, and Jesuit education in late Renaissance Italy.* Leiden; Boston, Brill.

Gunnell, D., Kidger, J., Elvidge, H., 2018. Adolescent mental health in crisis. *BMJ* 361, k2608. Access Number: 29921659.

Haber, S.N., 2016. Corticostriatal circuitry. *Dialogues in Clinical Neuroscience* 18(1), 7–21. PMC4826773. Access Number: 27069376.

Haber, S.N., Knutson, B. 2010. The reward circuit: linking primate anatomy and human imaging. *Neuropsychopharmacology* 35(1), 4–26.

Howes, O.D., Kapur, S., 2009. The dopamine hypothesis of schizophrenia: version III – the final common pathway. *Schizophrenia Bulletin* 35(3), 549–562. PMC2669582. Access Number: 19325164.

Insel, T.R., 2014. The NIMH Research Domain Criteria (RDoC) Project: precision medicine for psychiatry. *American Journal of Psychiatry* 171(4), 395–397. Access Number: 24687194.

Johnson, P.M., Kenny, P.J., 2010. Dopamine D2 receptors in addiction-like reward dysfunction and compulsive eating in obese rats. *Nature Neuroscience* 13(5), 635–641. PMC2947358. Access Number: 20348917.

Kandel, E.R. (2013). *Principles of neural science.* New York, McGraw-Hill.

Kemp, S., 2019. Mental disorder and mysticism in the late medieval world. *History of Psychology* 22(2), 149–162. Access Number: 30730170.

Kessler, R.C., Berglund, P., Demler, O., Jin, R., Merikangas, K.R., Walters, E.E., 2005. Lifetime prevalence and age-of-onset distributions of DSM-IV disorders in the National Comorbidity Survey Replication. *Archives of General Psychiatry* 62(6), 593–602. Access Number: 15939837.

King, D.R., Nanda, G., Stoddard, J., Dempsey, A., Hergert, S., Shore, J.H., Torous, J., 2023. An introduction to generative artificial intelligence in mental health care: considerations and guidance. *Current Psychiatry Reports* 25(12), 839–846. Access Number: 38032442.

Kirchberg, M.C., Pinson, C., Frank, G.K.W., 2024. Pharmacotherapeutic strategies for the treatment of anorexia nervosa – novel targets to break a vicious cycle. *Expert Opinion on Pharmacotherapy*, 1–13. Access Number: 39497232.

LeDoux, J., 2003. The emotional brain, fear, and the amygdala. *Cellular and Molecular Neurobiology* 23(4–5), 727–738. Access Number: 14514027.

McClellan France, J., Jovanovic, T., 2023. Human fear neurobiology reimagined: can brain-derived biotypes predict fear-based disorders after trauma? *Neuroscience and Biobehavioral Reviews* 144, 104988. PMC10960960. Access Number: 36470327.

Mora, G., 1980. Mind-body concepts in the Middle Ages: part II. The Moslem influence, the great theological systems, and cultural attitudes toward the mentally ill in the late Middle Ages. *Journal of the History of the Behavioral Sciences* 16(1), 58–72. Access Number: 6987295.

Nestler, E.J., Hyman, S.E. 2010. Animal models of neuropsychiatric disorders. *Nature Neuroscience* 13(10), 1161–1169.

Nusslock, R., Alloy, L.B., Brody, G.H., Miller, G.E., 2024. Annual research review: Neuroimmune network model of depression: a developmental perspective. *Journal of Child Psychology and Psychiatry and Allied Disciplines* 65(4), 538–567. Access Number: 38426610.

Olweus, D., 1979. Stability of aggressive reaction patterns in males: a review. *Psychological Bulletin* 86(4), 852–875. Access Number: 482487.

Osler, M.J., Brush, S.G., Brookes, J.S. (2019). *Scientific revolution.* Enclopedia Britannica.

Perdicoyianni-Paleologou, H., 2009. The vocabulary of madness from Homer to Hippocrates. Part 1: the verbal group of mualphaiotanuomicronmualphaiota. *History of Psychiatry* 20(79 Pt 3), 311–339. Access Number: 20030162.

Phillips, M.L., 2023. Building a neurobiological framework to elucidate neural mechanisms, aid early risk detection, and develop new treatments for bipolar disorder. *Psychiatry Research* 329, 115521. PMC10841719. Access Number: 37797440.

Porter, R. (2002). *Madness: a brief history*. Oxford; New York, Oxford University Press.

Quirk, G.J., Likhtik, E., Pelletier, J.G., Pare, D., 2003. Stimulation of medial prefrontal cortex decreases the responsiveness of central amygdala output neurons. *Journal of Neuroscience* 23(25), 8800–8807. Access Number: 14507980.

Raine, A., Phil, D., Stoddard, J., Bihrle, S., Buchsbaum, M., 1998. Prefrontal glucose deficits in murderers lacking psychosocial deprivation. *Neuropsychiatry, Neuropsychology, and Behavioral Neurology* 11(1), 1–7. Access Number: 9560822.

Reed, C.H., Bauer, E.E., Shoeman, A., Buhr, T.J., Clark, P.J., 2021. Acute stress exposure alters food-related brain monoaminergic profiles in a rat model of anorexia. *Journal of Nutrition* 151(12), 3617–3627. PMC8643607. Access Number: 34522956.

Richard, J.M., Berridge, K.C., 2011. Nucleus accumbens dopamine/glutamate interaction switches modes to generate desire versus dread: D(1) alone for appetitive eating but D(1) and D(2) together for fear. *Journal of Neuroscience* 31(36), 12866–12879. PMC3174486. Access Number: 21900565.

Richard, J.M., Berridge, K.C., 2013. Prefrontal cortex modulates desire and dread generated by nucleus accumbens glutamate disruption. *Biological Psychiatry* 73(4), 360–370. PMC3535675. Access Number: 22981656.

Robinson, L., Zhang, Z., Jia, T., Bobou, M., Roach, A., Campbell, I., Irish, M., Quinlan, E.B., Tay, N., Barker, E.D., Banaschewski, T., Bokde, A.L.W., Grigis, A., Garavan, H., Heinz, A., Ittermann, B., Martinot, J.L., Stringaris, A., Penttila, J., van Noort, B., Grimmer, Y., Martinot, M.P., Insensee, C., Becker, A., Nees, F., Orfanos, D.P., Paus, T., Poustka, L., Hohmann, S., Frohner, J.H., Smolka, M.N., Walter, H., Whelan, R., Schumann, G., Schmidt, U., Desrivieres, S., Consortium, I., 2020. Association of genetic and phenotypic assessments with onset of disordered eating behaviors and comorbid mental health problems among adolescents. *JAMA Netw Open* 3(12), e2026874. PMC7711322. Access Number: 33263759.

Rose, N., 2016. Neuroscience and the future for mental health? *Epidemiology and Psychiatric Sciences* 25(2), 95–100. PMC6998666. Access Number: 26234570.

Rusbridge, C., 2020. Neurobehavioral disorders: the corticolimbic system in health and disease. *Veterinary Clinics of North America: Small Animal Practice* 50(5), 1157–1181. Access Number: 32680665.

Sallam, H.N., 2010. Aristotle, godfather of evidence-based medicine. *Facts Views Vis Obgyn* 2(1), 11–19. PMC4154333. Access Number: 25206962.

Shorter, E. (1997). *A history of psychiatry: from the era of the asylum to the age of Prozac*. New York, John Wiley & Sons.

Singh, M.K., Gorelik, A.J., Stave, C., Gotlib, I.H., 2024. Genetics, epigenetics, and neurobiology of childhood-onset depression: an umbrella review. *Molecular Psychiatry* 29(3), 553–565. Access Number: 38102485.

Smith, A.J., Smith, L.A., 2016. Viral carcinogenesis. *Progress in Molecular Biology and Translational Science* 144, 121–168. Access Number: 27865457.

Sporns, O., 2022. The complex brain: connectivity, dynamics, information. *Trends in Cognitive Sciences* 26(12), 1066–1067. Access Number: 36207260.

Tondo, L., Vaccotto, P.A., Vazquez, G.H., 2018. [A brief historical approach about the concept of paranoia]. *Vertex* 29(137), 72–77. Access Number: 30605198.

Van Renynghe de Voxvrie, G., 1993. Flemish psychiatry from the middle ages to the 18th century. *Acta Psychiatrica Belgica* 93(2), 83–96. Access Number: 8036936.

Wang, X.J., Hu, H., Huang, C., Kennedy, H., Li, C.T., Logothetis, N., Lu, Z.L., Luo, Q., Poo, M.M., Tsao, D., Wu, S., Wu, Z., Zhang, X., Zhou, D., 2020. Computational neuroscience: a frontier of the 21(st) century. *National Science Review* 7(9), 1418–1422. PMC8288724. Access Number: 34691537.

Weider, S., Shott, M.E., Nguyen, T., Swindle, S., Pryor, T., Sternheim, L.C., Frank, G.K.W., 2024. Food avoidance and aversive goal value computation in anorexia nervosa. *Nutrients* 16(18). PMC11434691. Access Number: 39339714.

Welch, A.C., Zhang, J., Lyu, J., McMurray, M.S., Javitch, J.A., Kellendonk, C., Dulawa, S.C., 2021. Dopamine D2 receptor overexpression in the nucleus accumbens core induces robust weight loss during scheduled fasting selectively in female mice. *Molecular Psychiatry* 26(8), 3765–3777. Access Number: 31863019.

8

Introduction to Neuroethics

Sukumar Vijayaraghavan

There is nothing either good or bad but thinking makes it so.

William Shakespeare (Hamlet, Act II, scene 2)

8.1 Introduction

At the outset, let us define two inter-related terms, morality and ethics, as used in this chapter. Morality is our sense of right and wrong, reflective of values held by a community that are necessary for its well-being. Ethics is the philosophical, scientific, and practical examination of morality. It examines the meaning of morality and the justification of normative standards and moral judgements.

From time immemorial, humans have contemplated the ethical framework of their existence. Around the 2nd millennium BCE, Hindu philosophers proposed the concept of *dharma* that encompasses principles of morality that one should abide by and includes living a life of compassion and empathy. The idea recognizes that, as humans are unique in that they have the capacity to discriminate and choose, they tend to create conflicts or moral dilemmas (Srivastava et al., 2013). Striving to recognize and address these internal conflicts is part of the human condition and brings one closer to being in harmony with cosmic law.

Hellenistic views, at least in earlier periods, were eudiamonistic, that is, aimed at the pursuit of happiness. While it has been argued eudaimonism incorporates within itself the idea of morality, philosophers from Socrates to Aristotle didn't directly broach the subject, neither justifying nor prescribing rules for a moral life. It was left to the later stoic and epicurean philosophers to explicitly link virtue of character with the theme of a good life and the pursuit of happiness.

These themes were extended during the Age of Enlightenment. Philosophers like Jeremy Bentham and John Stuart Mill were 'hedonistic act consequentialists' and held the view that an act is moral if it maximizes pleasure (intrinsic good) and minimizes pain (intrinsic bad). This implies that they felt that an act by itself has no intrinsic moral weight to it but only acquires a moral character based on its consequence, for example, the claim

DOI: 10.1201/9781032717791-8

that the bombings of Hiroshima and Nagasaki were morally justifiable because they prevented a much greater loss of life that would have resulted from a ground invasion. This contradicted moral philosophers like Immanual Kant, who proposed a principle that he called 'Categorical Imperative' that we must follow even if our impulses are to the contrary. Moral law is universal and has the respect of all humans, even those who do not abide by it.

In a classic paper, Haidt (2001) proposed a social intuitionist model for moral behavior, arguing against rationalist models for our moral judgement. He postulated that moral reasoning is an *ex post facto* process, in an effort to justify and disseminate one's innate responses. As an example, he cites the following hypothetical scenario. Consider a brother and sister who have decided it would be fun to engage in sex during a vacation trip. They take precautions to avoid pregnancy. They enjoy the experience and feel it brings them closer to each other. The question, according to Haidt, is whether it was OK for them to engage in this behavior.

People generally subscribe to the immorality of the situation and often provide rational explanations, such as inbreeding is bad (not so here, as they took extra precautions to avoid impregnation) or that it is emotionally damaging to the participants (again, not true in this example). Often at the end of such reasoning, people resort to "I just know it is wrong" (Haidt, 2001; Haidt et al., 2000). This inability to reason out their position, or being 'morally dumbfounded', usually failed to alter their moral judgement (Haidt et al., 2000; Haidt and Hersh, 2001). According to Haidt's model, a person's intuition guides his/her moral judgement, which is rationalized *post hoc*. The reasoning then is part of social discourse and is directed at influencing the intuition and judgement of others. Haidt's model is consistent with previous psychological experiments that show that people make choices for reasons unknown to them and then justifying them with reasonable-sounding logic (e.g., see Nisbett and Wilson, 1977).

The idea that our moral judgements have an emotional basis (Green et al., 2001; Greene et al., 2004; Greene, 2007) also informs our idea of the 'moral brain' in our efforts to understand neural mechanisms that drive these behaviors. Greene et al. (2004) proposed a 'dual-process' hypothesis which states, as did Haidt, that moral decisions that are deontological (morally obligatory in a Kantian sense or in rule-based ethics, from the Greek word *deont*, meaning that which is binding) are driven by emotional responses, while utilitarian decisions (based on the greater good) involve genuine moral reasoning rather than rationalization of emotional responses (Greene, 2007).

Advances in neuroethics lend credence to Darwin's idea that the roots of human morality lie in the social instincts in other species (1872/1965). Thus, it is valid to begin examination of the neurobiology of ethics with the biology of animal social behavior.

8.2 Analogs of 'Ethical' Behavior in Animals

As philosophers struggled with morality and ethics, the shift in the field of ethics from philosophy to biology began with the emergence of sociobiology as a discipline and E.O. Wilson's idea of "The New Synthesis' (1975), which wrested control (at least partially) from philosophers in an effort to 'biologize' morality and ethics.

It can be argued that prosocial behaviors in animals, combined with human abilities to introspect, choose, and establish cultural and societal norms, gave rise to moral and ethical behavior. From an evolutionary perspective, ethics can be seen as an adaptation that allows cooperation between people and promotes social harmony.

When it refers to animals, terms like altruism and empathy are loosely defined and anthropomorphic. To truly depict what one observes in humans, one needs to presume a degree of intentionality, awareness of self and other, and awareness of consequences. However, prosocial behavior in animals, irrespective of terminology, can be reasonably assumed to reflect the biological underpinnings of moral behavior in humans.

De Waal (2003) proposed a Russian doll model of the evolution of empathy that can be extrapolated to moral behavior in general. At the core of this model is the perception-action mechanism, which induces similar states in the observer and the target. For example, a well-known behavior in humans is yawn contagion, which is seen in chimpanzees (Campbell and de Waal, 2011), in dogs (Romero et al., 2013), and even in birds (Gallup et al., 2015). Enveloping this inner doll are behaviors that reflect empathic concern and consolation, reflecting emotional contagion. In mice, pain response in one individual enhances the pain response of an observing conspecific (Langford et al., 2006), reminiscent of empathy in humans. The outermost layer of the doll involves cognitive processes and perspective taking, that is, putting oneself in the shoes of the other. This involves top-down control of the core behaviors (de Waal and Preston, 2017).

Thus, it appears that arguments made against the validity of true empathy or altruism in animals might be moot from biological and evolutionary perspectives. These animal behaviors could be argued as foundations of moral behavior in humans. While claims have been made that altruistic behaviors in animals are merely motivated by reciprocal gain, we need to distinguish the actual behaviors from their motivation. A mouse does not engage in behaviors based on evaluation of future favors or personal gain (de Waal and Preston, 2017).

Social insects like bees, ants, and termites exhibit altruism, that is, behaviors that clearly are against self-interest. Worker bees will devote their lives to nurturing the queen, which is the only one that can reproduce. Darwin was initially perplexed by altruistic behavior and thought it went against his idea of natural selection (1859), as it clearly incurs individual fitness costs for the benefits of others. He and others like William Hamilton modified his theory to include the idea of inclusive fitness, where individuals can act altruistically for the benefit of one's social group as a whole (Dugatkin, 2007).

African grey parrots will voluntarily help each other receive food rewards. A parrot will readily donate a reward token to its conspecific that could be exchanged for food (Brucks and von Bayern, 2019). Vampire bats will share food with other conspecifics (Carter and Wilkinson, 2013).

In Africa, many chimpanzees are maimed or killed in traps set by people hunting for bush meat. However, in Bossou, Guinea, incidences of maiming are rarer than in other regions. Ohashi and Matsuzawa (2011) found that in Bossou, male chimpanzees were observed to deactivate traps (at the risk of harming themselves), presumably to keep the colony safe. Similar behavior is observed in young male gorillas to protect their tribe from poachers (Than, 2012). A surprising behavior was observed recently in harvester ants, where a rescue operation was mounted to save a nest member trapped in a spider web (Kwapich and Holdobler, 2019). The result suggests that such prosocial behaviors might be an innate property of all fauna that live in cooperative colonies.

8.3 Neuroethics

The term neuroethics was originally coined by Anneliese Alma Pontius (1973) in an article criticizing the work of Zelazo et al. (1972) on their work on "walking" exercises for newborns who clearly do not possess the walking reflex. Her article dealt with the ethical aspects of the study using the neologism 'neuro-ethics' as a branch of bioethics. Carnford (1989) proposed the term 'neuroethicist' for a neurologist concerned with ethical issues in the field. In 1995, the bioethics committee of UNESCO (Vincent, 1995) emphasized the relation between neuroscience and ethics. However, it was only in 2002 that the field was formally recognized when the Dana Foundation organized the first international conference recognizing the field and providing a definition for neuroethics.

> Neuroethics is the study of ethical, legal, and social questions that arise when scientific findings about the brain are carried into medical practice, legal interpretations, health and social policy. . . . Neuroethics must examine how doctors, lawyers, judges, insurance companies, and politicians, as well as society at large, make use of these facts.
>
> **(Marcus, 2002)**

In 2002, Adina Roskies presented two subdivisions of the field, the first dealing with the neuroscience of ethical behavior and the second, a branch of bioethics, dealing with the ethical issues arising from advances in neurosciences.

8.4 Neuroscience of Ethics – Insights from Animal Studies

Advances in neuroscience have done away with the mind–brain dichotomy. It is now widely accepted that the mind is a product of the brain and its network. That being said, it is naïve to assume that there is a specific brain locus that confers these properties. While there is no dedicated *nucleus moralis or fasciculus ethicus* in the brain (Giordano et al., 2016), we can assume, with little controversy, that moral behaviors, like all behaviors, have bases in brain physiology.

Many animal studies have examined the physiological underpinnings of what can be construed as behaviors within the ambit of human ethics. For example, American prairie voles (*Microtus ocrogaster*) show highly paternal and monogamous attachments, whereas the montane vole (*Microtus montanus*) is very promiscuous and does not exhibit paternal behaviors towards their offspring. Thus, these voles have provided us with an ideal model for comparative studies for examining these behaviors. For example, it has been shown that these voles have differences in the distribution of arginine vasopressin (AVP) and oxytocin (OT) receptors. Prairie voles have higher densities of AVP $V_{1a}R$ receptors in the accessory olfactory bulb, diagonal band, and laterodorsal and paraventricular thalamus, while montane voles have greater expression of the receptors in the prefrontal cortex (Insel et al., 1994). Similarly, prairie voles have higher densities of OTRs in the nucleus accumbens and prefrontal cortex, areas that show little expression in the promiscuous species (Smeltzer et al., 2006). The level of AVP-immunoreactive fibers increased in male voles with cohabitation with its mate but not with an unfamiliar female (Bamshad et al., 1994).

A more direct approach to determine the effects of these peptides on pair bonding was done using pharmacological techniques. Intracerebroventricular (icv) administration of V1aR antagonists in the male prairie voles prevented partner preference 24 h after mating, while icv application of AVP caused pair bonding in the absence of mating (Winslow et al., 1993; Cho et al., 1999). Interestingly, icv injection of AVP had no effect on promiscuous voles (Young and Wang, 2004).

Examining the gene structure of $V_{1a}R$ and OTR underlying this life strategy showed that the demonstrated coding regions are conserved among the monogamous species and different in the promiscuous voles. For example, the monogamous voles have repetitive microsatellite DNA sequence in the promoter regions of the $V_{1a}R$, not present in the promiscuous species (Young et al., 1997). Transgenic mice that received the prairie vole $V_{1a}R$ gene showed receptor distribution different from that of wild-type mice and showed increased affiliative behavior (Young et al., 1999). However, it is important to note, as stated in the beginning of this section, these posterior pituitary hormones are not likely to be the only determinants of pair bonding behavior in rodents (e.g., see Fink et al., 2006). It is highly likely that other systems are involved as well. For example, injection of a dopaminergic (DA) agonist induced pair bonding in the absence of mating, while a non-specific DA receptor antagonist blocked the behavior upon mating in monogamous voles (Aragona et al., 2003). More work needs to be done to understand the mechanistic underpinnings of these social behaviors.

These studies imply that hormonal and neurotransmitter signals underlie social behaviors and have relevance to our ability to form social bonds and, by extension, codify rules that are aimed at maintaining them. In addition, these studies provide us with insights for the treatment of conditions like autism, schizophrenia, and social anxiety disorders, where the inability to form social bonds is a diagnostic feature (Volkmar, 2001).

8.5 Morality and the Human Brain

The roots of human morality are intricately linked to our social structure and attachments. Hence, the neurobiology of ethics is closely tied to neurochemical mechanisms underlying prosocial behaviors in humans as well. Humans and a few other species form long-term attachments within their social group. The neurobiology of these behaviors appear to be conserved across these species (Insell and Young, 2001). It has been suggested that these attachments are the drivers for the construction of our ethical systems (Berendzen, 2023).

In the last two decades, there have been a number of studies interrogating the neurobiological bases of human cognitive, emotional, and social behaviors, some of which fall under the ambit of moral actions (Darragh et al., 2015). Are there similar hormonal correlates for human moral decision making? This area of interrogation is new and mainly correlative.

Studies on the effects of hormones on human moral decision making can be divided in three groups: 1) studies where individuals are asked to respond to moral conundrums in the presence or absence of exogenously delivered hormones, 2) behavioral genetic studies where responses to moral dilemmas are correlated with genetic variants of hormonal receptors, and 3) fMRI studies examining brain activity under the same circumstances. Freitas and Osório (2022) published a metanalysis of literature showing association between the levels of different hormones and moral judgement.

Both reason and emotion are likely to play significant roles in moral decision making. Thus, these tasks are geared to tap into both such mechanisms. For example, a commonly used task is the trolley dilemma. Here an individual is asked to respond to the following scenario. A runaway trolley is headed on a collision course with five people in its path. A person has the option to divert the trolley by pulling a lever, which will divert it on a different course where there is only one person in its way. A more emotionally demanding variant of this dilemma is the footbridge dilemma, where one is standing on a footbridge with another person and the only way to prevent the runaway trolley from killing the five people is to push this person into its path, thus killing one to save five (Greene et al., 2001).

Using multiple tasks (20 non-moral, 25 personal moral, and 17 impersonal moral), Scheele et al. (2014) reported that exogenous delivery of oxytocin during moral decision-making tasks promoted self-interest moral judgements in men but a more altruistic response for the benefit of others in women. The self-centered responses in men might seem unexpected considering our previous discussions on the role of oxytocin. However, one way of interpreting these results is that this potentially selfish behavior in males might incorporate significant others with whom the person has close social bonds (Krienen et al., 2010). Thus, oxytocin might evoke a protective, prosocial response in males, consistent with its role in promoting social bonds.

Another effect of oxytocin appears to be to alter maladaptive rule adherence (Gross and De Dreu (2017). Given an arbitrary rule that demanded the participants forgo financial benefits, participants delivered exogenous oxytocin violated the rule more often, suggesting that oxytocin suppresses the need for a highly structured environment and promotes more flexibility in decision making, suggesting that the hormone shifts assessments from a deontologic to a more utilitarian mode.

Bernhard et al. (2016) showed that differences in OTR and AVPR genes are associated with differences in moral behavior. They analyzed single-nucleotide polymorphisms (SNPs) in the genes and peoples' ability to make moral judgements. A SNP (rs 237889) located at the intron of the OTR gene showed a significant correlation with frequency of utilitarian judgement in moral dilemma tasks with a linear trend across T/T, T/C, and C/C genotypes. In one study reported in the paper, a SNP (rs 1042615) in the AVPR gene also showed a similar correlation. Other OTR SNPs have been associated with prosocial temperament in humans (Tost et al., 2010) and empathy (Rodrigues et al., 2009).

Other hormones and neurotransmitters have been implicated in moral behavior (Freitas and Osório, 2022), like cortisol (Kossowska et al., 2016) and testosterone (Gong et al., 2017; Armbruster et al., 2021). It has been reported that increased stress, via cortisol, can drive intuitive and self-interest–driven moral decision making than the more cognitively demanding utilitarian ones (Starcke et al., 2012; Youssef et al., 2012). In contrast, androgen receptor SNPs (Gong et al., 2017), endogenous testosterone levels (Armbruster et al., 2021), and exogenous administration of the hormone (Chen et al., 2016) all seem to skew moral decision making towards more utilitarian responses, mainly in females.

Using fMRI, Shenhav and Greene (2014) examined the role of amygdala and the ventromedial prefrontal cortex (vmPFC) in moral judgement tasks. They demonstrated that the amygdala responds to emotional assessment of a situation, while the vmPFC appears to respond to judgements that generate a more integrated assessment. A met-analysis of fMRI studies on moral decision making (Garrigan et al., 2016) examined the responses to both moral response decisions (i.e., responses to a moral dilemma) as well as moral evaluations (evaluating the appropriateness of another's moral response). A general conclusion is that in both cases, activation was observed in the left middle temporal gyrus, left cingulate gyrus, and left medial frontal gyrus. Another region commonly implicated in moral decision-making is the ventromedial prefrontal cortex. Young and Saxe (2008) found that moral judgement tasks preferentially activated the right temporo-parietal junction compared to tasks that were nonmoral. Predictably, these are common areas for cognitive processing predicted by the theory of mind (that refers to an ability to attribute mental states to oneself and others; Premack and Woodruff, 1978). Schurz et al. (2020) presented a meta-analysis of networks involved in social cognition that the reader can refer to.

Interpretation of fMRI data can be problematic while interpreting moral judgements. Extrapolating from Haidt's model, the validity of these experiments would depend on how effectively the moral conundrums used in the tests reflect the subjects' emotional involvement. Hence, it is important to ensure methodological homogeneity across studies using moral dilemmas that would allow for reliable meta-analysis. For example, in fMRI studies, the timing of the BOLD signal across trials might vary with the length (word count) of the dilemmas presented or the expression style (Christensen and Gomila, 2012). It also depends on participants' perspectives, based on their personal histories. For example, a conundrum involving abortion would elicit different levels of emotions in participants for whom it is a theoretical construct versus those who know someone who has undergone abortion, which in turn would be completely different for a person who has had one. Thus, much more stringent efforts need to be exercised in these studies to account for multiple potential confounds.

Studies on patients with brain lesions have provided some insights into areas that might be involved in moral reasoning and decision making. The historic case of Phineas Gage is a case in point. Gage, after having a tamping iron go through his head, survived but displayed emotional instability, aggressive behavior, and impaired social behavior. Similar behaviors have been shown in war veterans with lesions in the vmPFC. Consistent with the previous discussion, Damasio (1994), upon examination of his skull, determined that the likely location of the lesion was the vmPFC (Grafman et al., 1996). Individuals with congenital malformation of the vmPFC show severe antisocial behavior and impaired moral judgements later in life (Taber-Thomas et al., 2014). Another informative source for examining the neural correlates of morality is studies on people with psychopathy (Seara-Cardoso, 2023). Psychopathy is a pathology characterized by a lack of guilt and empathy. Callous unempathic behavior is one of the hallmarks of psychopathy (Viding and Kimonis, 2018). From fMRI studies using moral dilemmas, these individuals show reduced activation of vmPFC and amygdala but greater activation of other areas, suggesting that they might use different, more cognitive areas in their judgements (Glenn et al., 2009a, 2009b).

All these studies show that human moral judgements engage both emotional and cognitive areas of the brain, and their interplay determines our ability to be moral and ethical individuals. By necessity, conclusions from these studies are correlative, and we know less about the mechanisms underlying these phenomena. However, as the field progresses, coupled with animal studies, we will develop a clearer understanding of how we make moral decisions.

8.6 Ethics of Neuroscience

While it is clear from the discussion is that neuroscience research can shed light on a number of issues regarding moral judgements and play a role in normative debates; it is also true that advances in neuroscience research can raise hitherto unknown moral conundrums. Advances in brain imaging techniques and drugs that enhance cognition can raise ethical questions. When is it justifiable to peer into someone's brain using fMRIs? Is it ethical to use cognitive enhancers to gain competitive advantage in a job interview?

Some see the ethics of neuroscience as a continuum in our exploration of bioethics starting from prefrontal lobotomy of the mentally ill, through the Nuremberg trials, Tuskegee syphilis experiments, to the current conundrums raised by neuroscience research (Iles and Bird, 2006). There has been some pushback against balkanization of the field of ethics considering "hyphenated ethics" (e.g., neuroethics, genethics) counterproductive, as it risks wasting time reinventing the wheel (Parens and Johnston, 2007). However, there is an argument to be made that to fully understand ethical issues raised by a field requires intricate understanding of said field. Therefore, such subdivisions are practical (Roskies, 2002) while recognizing, at the same time the need for a larger group of scholars to evaluate emerging issues for the benefit of society as a whole. As the neuroscientist Eric Kandel stated at a DANA Foundation meeting during the Society for Neuroscience annual meeting in 2005, "I am here tempted to paraphrase in the context of ethics what Reinhold Niebuhr, the great Columbia theologian, said about democracy: the people's capability for doing good makes biological ethics desirable; their capability for evil makes biological ethics necessary" (as reproduced in Illes and Bird, 2006).

There are two types of issues that underpin neuroethics. First, the conduct of neuroscience research itself raises ethical issues both with animal and human subjects. Second, advances in our understanding of the brain and the development of novel neurotechnologies provide hitherto unimaginable access to our brains, raising questions on what it means to be human. It is imperative that there be an ethical framework under which this work is conducted. In this section, we will consider ethical issues raised by neuroscience research and tackle one example issue raised by advances in neurotechnology and neuropharmacology: cognitive enhancement.

8.7 Ethics of Neuroscience Research

In 1962, Dr. Louis Jolyon West and his colleagues conducted the following experiment. To assess whether Lysergic Acid Diethylamide (LSD) induced the phenomenon of musth (a state of heightened testosterone-induced aggression) in elephants, they injected Tusko, a male resident of the Oklahoma zoo, with 297 mg acid (a huge dose for humans even when scaled up to the weight of an elephant). The elephant died, either from the acid or from the fact that they injected him later with Thorazine to calm him down. I use this as an example to illustrate the perils of unregulated research. There was no strong scientific or ethical justification for the experiment. In 1965, the Association for Assessment and Accreditation of Laboratory Animal Care International (AAALAC) was formed to foster the ethical treatment of animals in research. They extended the concept of replacement, reduction, and refinement (3Rs), originally proposed by Russell and Burch (1959). Russell and Burch noted (as reproduced in Tannenbaum and Bennett, 2015)

> We turn now to consideration of the ways in which inhumanity can be and is being diminished or removed. These ways can be discussed under the 3 broad headings of Replacement, Reduction, and Refinement. . . . [T]he 3 modes now considered have conveniently been referred to as the 3Rs of humane technique. Replacement means the substitution for conscious living higher animals of insentient material. Reduction means reduction in the numbers of animals used to obtain information of a given amount and precision. Refinement means any decrease in the incidence or severity of inhumane procedures applied to those animals which still have to be used.

While human research can be considered deontological, animal research is utilitarian (see Woollard, 2021 for detailed discussion; Rickless, 1997). It has been argued that research on non-human primates should shift towards approaches that adhere to deontological principles using the same criteria used for human subjects (Carvalho et al., 2019).

Animal research in neuroscience is justifiable by the enormous positive impact it has on humans. Cochlear implant studies on a number of mammalian animal models have benefitted thousands of deaf humans. Deep brain stimulation studies on non-human primates have yielded treatments for Parkinsonism in humans (Wichmann et al., 2018). At the same time, a vast number of brain disorders and mental illnesses still await effective treatments, resulting in individual suffering and distress and a huge economic burden on society. Animal research is likely to be the foundation for these disorders for many years to come.

Neuroscience research with animals often falls short of the 3 Rs by necessity. Brain research can often not be done *in silico* or even with cell lines and primary cell cultures. While reduced preparations like brain organoids that are derived from human-induced pluripotent stem cells (hIPSCs) have been used to limited extent, such as to screen drugs, they fall short of providing relevant physiological information in many cases (Homberg et al., 2021) The aim being to understand how the human brain works, there is often a phylogenetic cutoff below which models can become irrelevant, especially while studying the physiological bases of behavior. Using a utilitarian approach, it is imperative that we use the lowest rung in the phylogenetic tree that allows us to answer specific questions regarding the brain. It is our collective ethical responsibility to adhere to this principle. It is in the interest of all neuroscientists to take inordinate care in designing experiments with animals, especially when working with non-human primates. This includes transparency and strict attention to regulatory policies as well as public engagement to convey the need and the importance of their research (Mitchell et al., 2021).

If I started this section with an example of unethical use of animals in research, then human subjects haven't faired well either. From Josef Mengele's experiments to the Tuskegee syphilis study, there have been numerous instances of human research that would fall under the umbrella of dubious to categorically unethical behavior. The Tuskegee syphilis experiments ran until 1972, long after the Nuremberg code of ethics for human experiments was accepted and even then were terminated only after Peter Buxtun, a public health official, leaked the story to the press and there was public outcry.

In response, Congress passed the National Research Act in 1974 that created a commission for the protection of human subjects of biomedical and behavioral research. The ethical principles outlined by the commission were summarized by the Belmont Report in 1979. The report identified core ethical principles that need to be adhered to when conducting research using human subjects. Neuroscience research on humans raises additional ethical issues. The brain being intricately tied to our sense of self, we have a more heightened sense of safety and privacy. When it comes to brain manipulations for research, we need clear articulation of policies in the specific context of brain research.

Let us take the example of stem cell transplant for the treatment of Parkinson's disease (PD) to illustrate the complexity of ethics in human neuroscience research. In 1980, Bjorklund and his group demonstrated that cell suspensions from the substantia nigra could be grafted in rats pre-injected with 6-hydroxy dopamine to eliminate their resident neurons. These grafted cells could differentiate and form functional connections in the neostriatum (Bjorklund et al., 1980). Less than a decade later, human experiments were initiated. Transplant of fetal dopamine neurons in an individual with PD showed improvements in rigidity and bradykinesia (Lindvall et al., 1990). In the US, the use of embryonic tissue ran against the moral intuitions of the religious right. If one believes in the personhood of embryos, then these experiments appear unethical. Until 1993, federal policy did not allow the NIH to fund research on transplanting fetal dopamine neurons to humans. This ban was removed under President Clinton but reinstated in part by President Bush in 2001.

As with other neurosurgical manipulations, ethical issues arose with the randomized double-blind studies undertaken for fetal transplants in PD. While placebo controls for pharmacological studies are inherently acceptable, sham surgeries in transplantation studies generated significant controversy, being invasive. Patients in the control group had to be exposed to the risks of the surgery (drilling burr holes into a skull) without any potential benefit from the procedure. On the other hand, to reduce risk, the dura wasn't penetrated in control subjects unlike the test group, where it was unavoidable, leading to questions regarding the appropriateness of the control procedure for risk benefit analyses. In one study (Olanow et al., 2003), all groups received immunosuppressants for 18 months, exposing the sham participants to the risk of infections without any potential benefits.

Placebo effects are a problem in studies involving brain manipulations. In a metanalyses of open-label and double-blind studies with PD patients, Polgar and Mohamed (2019) concluded that sham surgery for PD is not necessary and therefore is unethical from a utilitarian standpoint. Their results suggested that the open-label trials resulted in a 33.7% improvement in symptoms, while in randomized control trials, the net effect (test-sham) was only 5.8%. This raises the question of whether results from placebo controls might be misleading and unnecessary in this context. Symptom changes in the sham controls were very variable. Indeed, in one study involving the use of fetal cell transplant in PD, it was observed that the benefit was better predicted not by actual but rather perceived treatment allocation (test vs. sham; McRae et al., 2004). One explanation could be that the placebo effect could be linked to reward mechanisms and result in the release of dopamine (De la Fuente-Fernandez et al., 2006) that would have a physiological benefit to the participants. On the other hand, placebo can also result in negative effects, called the nocebo effect, or a less effective one, called the lessebo effect (Mestre, 2020). By nature, all placebo-controlled studies are deceptive. There is an ethical conflict between beneficence and respect for autonomy (Stoessl, 2022).

The question of informed consent needs to be considered carefully in these experiments. Unfortunately, in an effort to ensure full disclosure, consent forms have become extremely long and complicated, often leading to confusion among participants. Also, many participants with PD may be cognitively compromised in a manner not easily determinable.

Though not comprehensive, the previous discussion raises ethical conflicts that can compromise human neuroscience experiments. As anticipated conditions can be varied depending on protocols, one easily implemented solution might be for institutional review boards to engage in much more in-depth dialogs with the investigators before approving protocols and not restrict themselves to checklists.

8.8 Cognitive Enhancement

We place a premium on the authenticity of our behaviors and are leery about the use of psychopharmacological manipulations and brain–computer interphases to improve our memory, attention, and performance. It is recognized that neuroscientists are primarily concerned with discovery and less with social implications of their findings, and thus, a broader coalition with social scientists is necessary to differentiate between what we can do and what we should do (Farah et al., 2004).

Cosmetic psychopharmacology as a term was first introduced by the author Peter Kramer (1997) in his popular book *Listening to Prozac* and can be defined as treating healthy people with psychotropic drugs to improve their performance and cognitive skills (Lugg, 2024). One common argument for the use of drugs to improve brain performance is that it is no different from physical enhancement, such as using botox or breast augmentation. In his biography of Steve Jobs, the founder of Apple, Isaacson stated the tech mogul attributed much of his entrepreneurial success to the use of LSD (Isaacson, 2011). Sigmund Freud famously promoted the use of cocaine as mental enhancement (Epstein and Farrell, 2019). The popular cosmologist Carl Sagan was an early advocate for the legalization of cannabis (Sagan, 1969). Indeed, it has been argued that the use of cognitive enhancers is as old as human history (McKenna, 1993).

It has been reported that up to 20% of young adults in this country use stimulants like Ritalin and Adderall to increase focus and performance (Kennedy, 2018), and these drugs are known to improve performance in healthy adults. In another study, about 41% of college students had used stimulants, mainly for help with studying (Garnier-Dykstra et al., 2012). Hence, it could be argued that the cat is out of the bag, and we need to develop processes to examine the safety, ethics, and governmental policies regarding the use of cosmetic drugs.

8.8.1 Brain-Computer Interfaces (BCIs) as Cognitive Enhancers

As with pharmacological interventions, let's begin by distinguishing the two uses of BCIs, first as clinical tools, for example, the use of deep brain stimulations to improve motor coordination in Parkinson's disease, inside-out interphases to move a prosthetic limb, or emerging technologies aimed at restoring vision (Karikari and Koshechkin, 2023), and second as cognitive enhancers in healthy individuals. We are mainly concerned with the latter.

With rapid advances in neurotechnology, we will be able to reliably decode brain activity. For example, relatively low-resolution fMRI images can be used to identify natural images based on brain activity (Kay et al., 2008). Another nascent area that will see rapid progress is brain–brain interfaces that will allow two people to communicate via their brain activity (Stocco et al., 2015). As technology improves to permit non-invasive recordings, the use of BCI to augment the intellectual and cognitive capabilities of healthy individuals will become prevalent (Goering and Yuste, 2016).

8.8.2 Ethical Issues

The libertarian view of this issue would be that people should be free to choose cognitive enhancers if they wish to. However, this is a bit simplistic. Taking our cues from anabolic steroid use in athletics, is it not up to society to determine the legality and validity of such use? If not, other issues emerge. Who gets to, or can afford to, take these drugs? Will we be creating an intellectual caste system based on income?

In the realm of cautionary tales, the ethics of cognitive enhancement is a flashback to Aldous Huxley's *Brave New World*. Unlike the other dystopian classic, George Orwell's *Nineteen Eighty Four*, which is a tale of your garden-variety authoritarian repression, in the Huxley classic, there is no coercion, just simple enticement. In *The New Republic Online* (June 21, 2001), the ethicist Leon Kass wrote that merely given the right technological imperatives, moral pluralism, and free markets, people in any liberal democracy can take themselves cheerfully into a 'Brave New World'. In his words "In case you had not noticed, the train has already left the station and is gathering speed, but no one seems to be in charge."

Our knowledge of brain physiology and psychopharmacology is not sufficient to be able to predict long-term consequences of the use of cognitive enhancers. There are very few, if any, longitudinal studies that can demonstrate the safety of these drugs over one's lifespan. One could argue that this is true for many psychotropic drugs used to treat mental illness, but there is a significant difference. Drugs prescribed for mental illnesses target debilitating symptoms that affect a person's normal functioning in the present. The same cannot be said for cognitive enhancer use in healthy adults who function normally without drugs.

Once the use of cognitive enhancement is normalized in society, do we run the risk of a cognitive arms race as newer and better drugs are developed or as BCIs become more efficient and affordable? There will always be a strong incentive for the pharmaceutical and technology companies to keep the golden goose alive and develop newer and better products in their race to capture larger market shares. This will add to the risk of introduction of agents whose long-term safety is less well documented, adding to the risk of brain dysfunction in later years. In the realm of long-term consequences, there is also no evidence that changes in cognition elicited by external manipulation recapitulate brain processes. For example, learning something difficult on our own has other effects on our brain/personality, such as capacity for hard work and self-discipline, that have long-term effects on other aspects of our lives.

There are other ethical issues underlying the treatment of children. Until recently, there was a distinct reluctance among psychiatrists to use psychoactive drugs in children diagnosed with mental illness. This has changed with the availability of 'safer' drugs, raising new ethical questions. For example, are the diagnostic criteria sound as delineated in the *Diagnostic and Statistical Manual of Mental Disorders*, or are they being influenced (or coerced) by teachers and parents to enforce appropriate behavior or to improve academic performance? Are we sculpting the natural development of the brain; in other words, are we moving towards creating designer children and less intellectual diversity in our adult population?

Last, there is the issue of coercion in adults. Could an employer insist workers take cognitive enhancers to improve efficiency as a condition for continued employment? Would it be okay if an employer mandates the use of brain–brain interfaces to increase the collective efficiency of the workforce? What implications does that have for individual creativity and aptitude, as well as our sense of self?

Unregulated practice on human subjects gave us the Tuskegee syphilis study and led to the formulation of the Belmont report (see previously). Currently, we lack any agreed-upon set of principles that should govern the use of cognitive enhancement, either with pharmacological agents or with BCI. There is a need to develop careful ethical guidelines for their use in healthy individuals.

8.9 Conclusion

It is not controversial to say that innovations in neuroscience are rapidly outpacing our society's ability to fully comprehend their ramifications. Our abilities to peer into brains, decode brain activities, and use a computer as an extension of our mind were the stuff of science fiction just a few short years ago. What can be lost in this excitement are negative consequences of these advances. In principle, everyone should be on board with the idea that progress in neuroscience ought to target the betterment of mankind. However, history has shown us that either due to lack of imagination or presence of malicious intent, the course of human innovation can take horrific turns. Neuroethics, as a discipline, provides the guardrails within which both neuroscience research and the use of neurotechnology and neuropharmacology should exist. Beyond biology, neuroscience has the potential to alter our view of humanity and our sense of self. Thus, the need for neuroethics to keep up with 'neuro-advancement' is urgent. As the Irish poet William Butler Yeats wrote in his poem "Aedh Wishes for the Cloths of Heaven," "Tread softly because you tread on my dreams."

References

Aragona BJ, Liu Y, Curtis JT, Stephan FK, Wang Z (2003). A critical role for nucleus accumbens dopamine in partner-preference formation in male prairie voles. *J Neurosci*. 23, 3483–3490.

Armbruster D, Kirschbaum C, Strobel A (2021). Androgenic morality? Associations of sex, oral contraceptive use and basal testosterone levels with moral decision making. *Behav Brain Res*. 408, 113196.

Bamshad M, Novak MA, de Vries GJ (1994). Cohabitation alters vasopressin innervation and paternal behavior in prairie voles (*Microtus ochrogaster*). *Physiol Behav*. 56, 751–758.

Berendzen KM (2023). Understanding social attachments as a window into the neural basis of prosocial behavior. *Front Neurol*. 10, 3389.

Bernhard RM, Chaponis J, Siburian R, Gallagher P, Ransohoff K, Wikler D, Perlis, RH, Greene JD (2016). Variation in the oxytocin receptor gene (OXTR) is associated with differences in moral judgment. *Soc Cogn Affect Neurosci*. 11(12), 1872–1881.

Björklund A, Schmidt RH, Stenevi U (1980). Functional reinnervation of the neostriatum in the adult rat by use of intraparenchymal grafting of dissociated cell suspensions from the substantia nigra. *Cell Tissue Res*. 212, 39–45.

Brucks D, von Bayern, AMP (2019). Parrots voluntarily help each other obtain food rewards. *Curr Biol*. 30, 292–297.

Campbell MW, de Waal FBM (2011). Ingroup-outgroup bias in contagious yawning by chimpanzees supports link to empathy. *PLoS ONE*. 6, e18283.

Carter GG, Wilkinson GS (2013). Food sharing in vampire bats: Reciprocal help predicts donations more than relatedness or harassment. *Proc R Soc B*. 280, 20122573.

Carvalho C, Gaspar A, Knight A, Vicente L (2019). Ethical and scientific pitfalls concerning research with non-human primates, and possible solutions. *Animals*. 9(12), 1–17.

Chen C, Decety J, Huang PC, Chen CY, Cheng Y (2016). Testosterone administration in females modulates moral judgment and patterns of brain activation and functional connectivity. *Hum Brain Mapp*. 37(10), 3417–3430.

Cho MM, DeVries AC, Williams JR, Carter CS (1999). The effects of oxytocin and vasopressin on partner preferences in male and female prairie voles (*Microtus ochrogaster*). *Behav Neurosci*. 113, 1071–1079.

Christensen JF, Gomila A (2012). Moral dilemmas in cognitive neuroscience of moral decision-making. *Neurosci Biobehav Rev*. 36, 1249–1264.

Cranford RE (1989). The neurologist as ethics consultant and as a member of the institutional ethics committee. The neuroethicist. *Neurol Clin*. 7, 697–713.

Damasio AR (1994). *Descartes' error: Emotion, reason, and the human brain*. New York: G.P. Putnam.

Darragh M, Buniak L, Giordano J (2015). A four-part working bibliography of neuroethics: part 2 – neuroscientific studies of morality and ethics. *PEHM*. 10(2), 1–22.
Darwin C (1859). *On the origin of species*. London: J. Murray, p. 204.
Darwin C (1872/1965). *The expression of emotions in man and animals*. Chicago: University of Chicago Press.
De la Fuente-Fernandez R, Lidstone S, Stoessl RJ (2006). Placebo effect and dopamine release. *J Neursal Transm Suppl*. 70, 415–418.
de Waal FBM (2003). *Feelings & emotions: The Amsterdam symposium* (eds. Manstead T, Frijda N, Fischer A). Cambridge Univ. Press, pp. 379–399.
de Waal FBM, Preston SD (2017). Mammalian empathy: behavioral manifestation and neural basis. *Nature Rev*. 18, 498–508.
Dugatkin LA (2007). Inclusive fitness theory from Darwin to Hamilton. *Genetics*. 176, 1375–1380.
Epstein K, Farrell HM (2019). "Miracle cures" in psychiatry. *Curr Psychiatry*. 18, 13–16.
Farah MJ, Illes J, Cook-Deegan R, Gardner H, Kandel E, King P, Parens E, Sahakian B, Wolpe PR (2004). Neurocognitive enhancement: What can we do and what should we do? *Nat Rev Neurosci*. 5, 421–425.
Fink S, Excoffier L, Heckel G (2006). Mammalian monogamy is not controlled by a single gene. *Proc Natl Acad Sci USA*. 103, 10956–10960.
Freitas CCMdC, Oso'rio FdL (2022). Moral judgment and hormones: A systematic literature review. *PLoS ONE*. 17(4), e0265693. https://doi.org/10.1371/journal.pone.0265693
Gallup AC, Swartwood L, Militello J, Sackett S (2015). Experimental evidence of contagious yawning in budgerigars (*Melopsittacus undulatus*). *Anim Cogn*. 18, 1051–1058.
Garnier-Dykstra LM, Caldeira KM, Vincent KB, O'Grady KE, Arria AM (2012). Nonmedical use of prescription stimulants during college: Four-year trends in exposure opportunity, use, motives, and sources. *J Am Coll Health*. 60, 226–234.
Garrigan B, Adlam ALR, Langdon PE (2016). The neural correlates of moral decision-making: A systematic review and meta-analysis of moral evaluations and response decision judgements. *Brain Cogn*. 108, 88–97.
Giordano J, Becker K, Shook JR (2016). On the "Neuroscience of Ethics" – Approaching the neuroethical literature as a rational discourse on putative neural processes of moral cognition and behavior. *J Neurol Neuromed*. 1, 32–36.
Glenn AL, Raine A, Schug RA (2009a). The neural correlates of moral decision-making in psychopathy. *Mol Psychiatry* 14, 5–6.
Glenn AL, Raine A, Schug RA, Young L, Hauser M (2009b). Increased DLPFC activity during moral decision-making in psychopathy. *Mol Psychiatry*, 14, 909–911.
Goering S, Yuste R (2016). On the necessity of ethical guidelines for new neurotechnologies. *Cell*, 167, 882–885.
Gong P, Fang P, Yang X, Ru W, Wang B, Gao X, Liu J (2017). The CAG polymorphism in androgen receptor (AR) gene impacts the moral permissibility of harmful behavior in females. *Psychoneuroendocrinology*. 80, 74–79.
Grafman J, Schwab K, Warden D, Pridgen A, Brown HR, Salazar AM (1996). Frontal lobe injuries, violence, and aggression: A report of the Vietnam head injury study. *Neurology*. 46, 1231–1238.
Green JD, Sommerville RB, Nystrom LE, Darley JM, Cohen JD (2001). An fMRI investigation of emotional engagement in moral judgement. *Science* 293, 2105–2108.
Greene JD (2007). The secret joke of Kant's soul. In: *Moral psychology* (ed. Sinnott-Armstrong W). MIT Press, pp. 35–79.
Greene JD, Sommerville RB, Nystrom LE, Darley JM, Cohen JD (2004). The neural basis of cognitive conflict and control in moral judgement. *Neuron*. 44, 389–400.

Gross J, De Dreu CKW (2017). Oxytocin conditions trait-based rule adherence. *Social Cog Affect Neurosci*. 12, 427–435.

Haidt J (2001). The emotional dog and its rational tail: A social intuitionist approach to moral judgement. *Psychol Rev*. 108, 814–834.

Haidt J, Bjorklund F, Murphy S. (2000). Moral dumbfounding: When intuition finds no reason. *Lund Psychol Rep*. 2, 1–23.

Haidt J, Hersh M (2001). Sexual morality: The cultures and reasons of liberals and conservatives. *J Appl Soc Psychol*. 31, 191–221.

Homberg JR, Adan RAH, Alenina N, Asiminas A, Bader M, Becker T, Begg DP, Blokland A, Burger ME, van Dijk G, Eisel ULM, Elgersma Y, Englitz B, Fernandez-Ruiz A, Fitzsimons CP, van Dam A-M, Gass P, Grandjean J, Havekes R, Henckens MJAG, Herden C, Hut RA, Jarrett W, Jeffrey K, Jezova D, Kalsbeek A, Kamermans M, Kas MJ, Kasri NN, Kiliaan AJ, Kolk SM, Korosi A, Korte SM, Kozicz T, Kushner SA, Leech K, Lesch K-P, Lesscher H, Lucassen PJ, Luthi A, Ma L, Mallien AS, Meerlo P, Mejias JF, Meye FJ, Mitchell AJ, Mul JD, Olcese U, Gonzalez AO, Olivier JDA, Pasqualetti M, Pennartz CMA, Popik P, Prickaerts J, de la Prida, LM, Ribeiro S, Roozendaal B, Rossato JL, Salari A-A, Schoemaker RG, Smit AB, Vanderschuren LJMJ, Takeuchi T, van der Veen R, Smidt MP, Vyazovskiy VV, Weismann M, Wierenga CJ, Williams B, Wiluhn I, Wohr M, Wolverkamp M, van de Zee EA, Genzel S (2021). The continued need for animals to advance brain research. *Neuron*. 109, 2374–2379.

Hume D (1896). *A treatise of human nature* (Original work in 1739). Oxford: Clarendon Press, p. 217.

Illes J, Bird SJ (2006). Neuroethics: A modern context for ethics in neuroscience. *Trends Neurosci*. 29, 511–517.

Insel TR, Wang ZX, Ferris CF (1994). Patterns of brain vasopressin receptor distribution associated with social organization in microtine rodents. *J Neurosci*. 14, 5381–5392.

Insel TR, Young LJ (2001). The neurobiology of attachment. *Nat Rev Neurosci*. 2, 129–136.

Isaacson W (2011). *Steve jobs*. New York: Simon & Schuster.

Karikari E, Koshechkin KA (2023). Review on brain-computer interphase technologies in healthcare. *Biophys Rev*. 15, 1351–1358.

Kay KN, Naselaris T, Prenger RJ, Gallant JL (2008). Identifying natural images from brain activity. *Nature*. 452, 352–355.

Kennedy S (2018). Raising awareness about prescription and stimulant abuse in college students through on-campus community involvement projects. *J Undergrad Neurosci Educ*. 17, A50–A53.

Kossowska M, Czernatowicz-Kukuczka A, Szumowska E, Czarna A (2016). Cortisol and moral decisions among young men: The moderating role of motivation toward closure. *Pers Individ Dif*. 101, 249–253.

Kramer P (1997). *Listening to Prozac: The landmark book about antidepressants and the remaking of the self*. New York: Penguin Books.

Krienen FM, Tu PC, Buckner RL (2010). Clan mentality: Evidence that the medial prefrontal cortex responds to close others. *J Neurosci*. 30, 13906–13915.

Kwapich CL, Holldobler B (2019). Destruction of spider webs and rescue of ensnared nestmates by a granivorous desert ant (*Veromessor pargandei*). *Am Nat*. 194, 395–404.

Langford DJ, et al. (2006). Social modulation of pain as evidence for empathy in mice. *Science*. 312, 1967–1970.

Lindvall O, Brundin P, Widnet H, Rehncrona S, Gustavil B, Frackowiak R, Leender KL, Sawle G, Rothwell JC, Marsden CD, Bjorklund MB (1990). Grafts of fetal dopamine neurons survive and improve motor function in Parkinson's disease. *Science*. 247, 574–578.

Lugg W (2024). Cosmetic psychiatry: A concept in urgent need of consideration. *Australas Psychiatry*. 32, 32–37.

Marcus SJ (2002). Neuroethics: Mapping the field. In: *Conference proceedings* (ed. Marcus SJ). New York: The Dana Press.

McKenna T (1993). *Food of the gods: The search for the original tree of knowledge*. Bantam Doubleday Dell Publishing Group Inc.

McRae C, Cherin E, Yamazaki TG, Diem G, Vo AH, Russell D, Ellgring JH, Fahn S, Greene P, Dillon S, Winfield H, Bjugstad KB, Free CR (2004). Effects of perceived treatment on quality of life and medical outcomes in a double-blind placebo surgery trial. *Arch Gen Psychiatry*. 61, 412–420.

Mestre TA (2020). Nocebo and lessebo effects. *Int Rev Neurol*. 153, 121–146.

Mitchell AS, Hartig R, Basso MA, Jarrett W, Kastner S (2021). International primate neuroscience research regulation, public engagement and transparency opportunities. *Neuroimage*. 229, 117700.

Nisbett RE, Wilson TD (1977). The Halo Effect: Evidence for unconscious alteration of judgements. *J Pers Soc Psychol*. 35, 250–256.

Ohashi G, Matsuzawa T (2011). Deactivation of snares by wild chimpanzees. *Primate*. 53, 1–5.

Olanow CW, Goetz CG, Kordower JF, Soessl AJ, Sossl V, Brin MF, Shannon KM, Nauert GM, Perl DP, Godbold J, Freeman TB (2003). A double-blind controlled trial of bilateral fetal nigral transplantation in Parkinson's disease. *Ann Neurol*. 403–414.

Parens E, Johnston J (2007). Does it make sense to speak of neuroethics? *EMBO Rep*. 8, S61–S64.

Polgar S, Mohamed S (2019). Evidence-based evaluation of the ethics of sham surgery for Parkinson's disease. *J Parkinsons Dis*. 9, 565–574.

Pontius AA (1973). Neuro-ethics of "walking" in the newborn. *Percept Mot Skills*. 37, 235–245.

Premack D, Woodruff G (1978). Does the chimpanzee have a theory of mind? *Behav Brain Sci*. 4, 515–526.

Rickless SC (1997). The doctrine of doing and allowing. *Philos Rev*. 106, 555–575.

Rodrigues SM, Saslow LR, Garcia N, John OP, Keltner D (2009). Oxytocin receptor genetic variation relates to empathy and stress reactivity in humans. *Proc Natl Acad Sci*. 106(50), 21437–21441.

Romero T, Konno A, Hasegawa T (2013). Familiarity bias and physiological responses in contagious yawning by dogs support link to empathy. *PLoS ONE*. 8, e71365.

Roskies A (2002). Neuroethics for the new millennium. *Neuron*. 35, 21–23.

Russell WMS, Burch RL (1959) (as reprinted 1992). *The principles of humane experimental technique*. Wheathampstead, UK: Universities Federation for Animal Welfare.

Sagan CMX (1969). *Marihuana reconsidered*. Harvard University Press.

Scheele D, Striepens N, Kendrick KM, Schwering C, Noelle J, Wille A, Schlspfer TE, Maier W, Hurlemann R (2014). Opposing effects of oxytocin on moral judgement in males and females. *Hum Brain Mapp*. 35, 6067–5076.

Schurz M, Maliske L, Kanske P (2020). Cross-network interaction in social cognition: A review of findings on task related brain activation and connectivity. *Cortex*. 130, 142–157.

Seara-Cardoso A (2023). The "(a)moral brain": When things go wrong. *Handb Clin Neurol*. 197, 107–117.

Shenhav A, Greene JD (2014). Integrative moral judgement: Dissociating the roles of amygdala and the ventromedial prefrontal cortex. *J Neurosci*. 34, 4741–4749.

Smeltzer MD, Curtis JT, Aragona BJ, Wang Z (2006). Dopamine, oxytocin, and vasopressin receptor binding in the medial prefrontal cortex of monogamous and promiscuous voles. *Neurosci Lett*. 394, 146–151.

Srivastava C, Dhingra V, Bharadwaj A, Srivastava A (2013). Morality and moral development: Traditional Hindu concepts. *Indian J Psychiatry*. 55, 283–287.

Starcke K, Ludwig A-C, Brand M (2012). Anticipatory stress interferes with moral judgement. *Judgem Decis Mak*. 7, 61–68.

Stocco A, Prat CS, Losey DB, Cronin JA, Wu J, Abernethy JA, Rao RPN (2015). Playing 20 questions with the mind: Collaborative problem solving by humans using a brain-brain interface. *PLOS One* 10, e01373303.

Stoessl AJ (2022). Deception and the ethics of placebo. *Int Rev Neurobiol*. 153, 147–163.

Taber-Thomas BC, Asp EW, Koenigs M, Sutterer M, Anderson SW, Tranel D (2014). Arrested development: Early prefrontal lesions impair the maturation of moral judgement. *Brain*. 137, 1254–1261.

Tannenbaum J, Bennett BT (2015). Russell and Burch's 3Rs then and now: The need for clarity in definition and purpose. *J Am Assoc Lab Anim Sci*. 54, 120–132.

Than K (2012). Gorilla youngsters seen dismantling poachers traps. National Geographic Online. *National Geographic Society*. http://news.nationalgeographic.com/news/2012/07/120719-young-gorillas-juvenile-traps-snares-rwanda-science-fossey/.

Tost H, Kolachana B, Hakimi S, Lemaitre H, Verchinski BA, Mattay VS, Weinberger DR, Meyer-Lindenberg A (2010). A common allele in the oxytocin receptor gene (OXTR) impacts prosocial temperament and human hypothalamic-limbic structure and function. *Proc Natl Acad Sci*. 107(125, 419–430.), 13936–13941.

Viding E, Kimonis ER (2018). Callous-unemotional traits. In: *Handbook of psychopathy*. The Guilford Press, pp. 144–164.

Vincent JD (1995). Ethics and neurosciences. In: *UNESCO. Proceedings. Third Session*. Paris: International Bioethics Committee of UNESCO, pp. 1–8.

Volkmar FR (2001). Pharmacological interventions in autism: theoretical and practical issues. *J Clin Child Psychol*. 30, 80–87.

Wichmann T, Bergmann H, Delong MR (2018). Basal ganglia movement disorders and deep brain stimulation: Advances through non-human primate research. *J Neural Transm*. 125, 419–430.

Wilson EO (1975). *Sociobiology: The new synthesis*. Belknap Press, of Harvard University Press.

Winslow JT, Hastings N, Carter CS, Harbaugh CR, Insel TR (1993). A role for central vasopressin in pair bonding in monogamous prairie voles. *Nature*. 365, 545–548.

Woollard F (2021). 'Utilitarianism for animals: deontology for people' and the doing/allowing distinction. *Philos Stud*. 180, 1149–1168.

Young L, Saxe R (2008). An fMRI investigation of spontaneous mental state inference for moral judgement. *J Cog Neurosci*. 21, 1396–1405.

Young LJ, Nilsen R, Waymire KG, MacGregor GR, Insel TR (1999). Increased affiliative response to vasopressin in mice expressing the V1a receptor from a monogamous vole. *Nature*. 400, 766–768.

Young LJ, Wang Z (2004). The neurobiology of pair bonding. *Nat Neurosci*. 7, 1048–1054.

Young LJ, Waymire KG, Nilsen R, Macgregor GR, Wang Z, Insel TR (1997). The 5′ flanking region of the monogamous prairie vole oxytocin receptor gene directs tissue-specific expression in transgenic mice. *Ann N Y Acad Sci*. 807, 514–517.

Youssef FF, Dookeeram K, Basdeo V, Francis E, Doman M, Mamed D, Maloo S, Degannes J, Dobo L, Ditshotlo P, Legall G (2012). Stress alters personal moral decision making. *Psychoneuroendocrinology*. 37(4), 491–498.

Zelazo PR, Zelazo NA, Kolb S (1972). "Walking" in the newborn. *Science*. 176, 314–315.

9

The Capacity for Evaluation of the Human Brain and Its Implications for an Artificial Moral Subject

M. Farisco and K. Evers

9.1 Introduction: The Ethics of AI and the Question of Artificial Moral Agency

Very broadly, ethics attempts to systematize and justify concepts of right and wrong and to clarify the implications of these concepts for behaviour and social rules. The ethical discussion on AI has ancient roots (Wiener 1954) and has developed alongside technical progress (Dreyfus 1972; Whitby 1991) to eventually become central in contemporary debates about AI (e.g., machine-learning and generative AI). Both the approaches to AI ethics and the specific topics considered vary quite widely (Dignum 2019; Coeckelbergh 2020; Stahl 2021), spanning privacy and data protection (Kaplan and Haenlein 2019; EDPS 2020); the need for AI transparency and explainability (Yeung 2018; Friedrich, Mason et al. 2022); the risk of biases and discrimination (Team 2018; Birhane 2021); and the potential societal impact of AI (Zuboff 2019; Walton and Nayak 2021), including its political (Nemitz 2018; Parliament 2020) and environmental impacts (Nishant, Kennedy et al. 2020), among others. In fact, the specific topics arising from AI depend on two factors in particular: the sector where AI is used and the background understanding of AI.

As recently proposed with reference to brain-inspired AI (Farisco, Baldassarre et al. 2024), it is possible and useful to distinguish between fundamental/foundational and practical ethical issues arising from AI: both relevant values and principles and key notions like moral subject, moral reasoning, and moral action, among others, including their implications for decision making, are key to justify moral choices. Accordingly, two main kinds of ethical issues arise from AI: foundational (i.e., concerning both the reasons for supporting research and investment in AI with the specific goal to make AI "moral", and the impact of AI on how we think about fundamental moral notions) and practical (e.g., concerning the impact of AI on our daily life). The first kind of issues involves theoretical analysis, while the second kind of issues involves mainly applied analysis, that is, the use of ethical theory to identify and address the practical issues related to the use of AI.

DOI: 10.1201/9781032717791-9

Farisco et al. (2024) propose that the practical issues arising from AI can be organized in terms of (at least) the following main levels:

- *Operational*, related to how AI works;
- *Instrumental*, related to how people use AI;
- *Relational*, related to how people see AI and the resulting psychological and metaphysical human–AI relationship; and
- *Societal*, related to the social and economic costs and consequences of the development and use of AI.

Also, Farisco et al. (2024) identify two main categories of fundamental ethical issues:

- *Those related to goals*, which refer to questions like: What is the driver of AI? What do we want to achieve by it?
- *Those related to concepts*, which include issues like implicit assumptions and biases about AI and considerations about the historical, cultural, and societal contexts of AI.

Thus, the discussion is open regarding both the methodology and the scope of AI ethics (Farisco, Evers et al. 2020). Importantly, in addition to the ethical issues related to the different possible uses of AI and related implications, there are ethical issues related to the possible understanding of AI as an ethical agent or a moral subject (i.e., capable of making moral choices). Presently, considering or defining AI as a possible ethical agent or a moral subject is highly contested (Misselhorn 2022).

In this chapter, we touch on this point through the analysis of a specific capacity that arguably plays a crucial role for qualifying humans as moral subjects: the capacity for evaluation, that is, the ability to discriminate between good and bad things in the world, assigning them positive or negative values on the basis of specific needs, motivations, and goals. Evaluation, and the related concept of value, is arguably the ground for making possible a number of cognitive capacities, including ethical deliberation.

In the following, we first provide a description of what evaluation is, with specific reference to its cognitive and emotional dimensions; then we focus on the role of evaluation for ethical deliberation specifically, and we conclude with some reflections on an illustrative model of artificial evaluation, which, according to the developers, promises to inform the artificial system with a capacity for moral reasoning. We conclude by arguing that this last point is conceptually not justified.

9.2 What Does Evaluation Mean?

Evaluation may be broadly described as the capacity for assessing the valence of processed information, which may originate either from the inside or from the outside of the organism/system. Therefore, evaluation results from the ability to discriminate between good and bad things in the world on the basis of specific needs, motivations, and goals. In biological organisms, this ability relies on the sensitivity to reward signals. Accordingly, to evaluate is more than perceiving. In order to evaluate a stimulus, a top-down feedback process is necessary: in fact, the processed stimulus is assessed on the basis of background priors which make it possible to evaluate its relevance and salience. In other words, not all stimuli are equally important for the experiencing subject, and their differences depend on a variety of background factors (e.g., past experiences, needs, goals, expectations, etc.). As such, evaluation is a defining feature of living organisms (Bennett 2023) and more specifically of the human brain, particularly of cognition, including conscious cognition. In fact, learning something may be facilitated or even made possible by assessing its valence (i.e., positive or negative value) for the subject. Without any value, or any capacity to evaluate stimuli, a system can neither learn or remember: it has to prefer some stimuli to others in order to learn. This classical idea in learning theory has been expressed in neuronal terms by, notably, Edelman, Changeux, and Dehaene in theories of consciousness and knowledge acquisition and later in models of predictive coding developed by, notably, Friston, suggesting that the brain is constantly generating and updating hypotheses that predict sensory input at varying levels of abstraction. In these accounts, learning is a change of behaviour that results from cerebral categorizations of stimuli in terms of positive or negative values, and value systems are conceived to be essential to the selectional workings of the brain. Evers suggested that this evaluative, self-projective, and predictive mode of operations of the human brain is the "evolutionary hallmark" of consciousness, which is a figurative way of saying that the former signifies the latter, that the evolution of preferences makes consciousness possible: they are, so to speak, the seal of consciousness (Evers 2009). They are also the foundation of human moral thought and judgment.

Importantly, evaluation and the related concept of value include both cognitive and emotional dimensions: in fact, cognitive representations like concepts, goals, and beliefs are combined with emotional attitudes having positive or negative valence (https://www.psychologytoday.com/intl/blog/hot-thought/201304/what-are-values). Emotions are particularly crucial for evaluating information coming from the inside of the organism (Bennett 2023).

Even if we share a basic intuitive understanding of emotion, there is no agreed-upon definition of it (LeDoux 1998; Lane, Nadel et al. 2000; Pessoa 2009; Adolphs and Anderson 2018). In fact, some include drive and motivation as definitional components of emotions, understood as states elicited by rewards and punishers (Rolls 2005), while others explicitly refer emotions to the evaluation of events (Arnold 1960) (i.e., appraisals). Also, the proposed taxonomies of emotions vary quite widely, spanning from so-called basic emotions (Ekman 1992) (e.g., fear, anger) to moral emotions (Haidt 2003; Moll et al. 2005) (e.g., pride, envy). Importantly, body and embodiment arguably play an important role for emotions (Damasio 1994).

For the sake of the present analysis, we may assume emotion as the functional link between an organism's expectation and actual stimuli mediated by specific structural and physiological mechanisms working as a connection between the internal and the

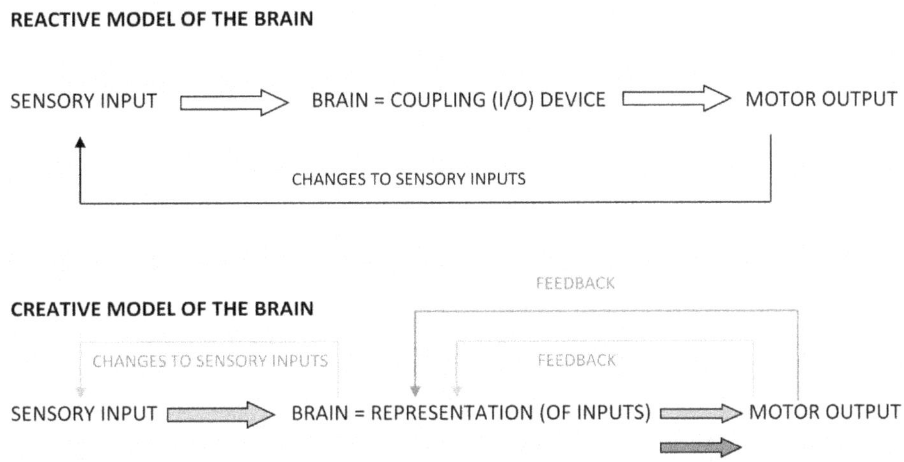

Figure 9.1 Reactive vs. creative model of the brain.

In the reactive model, the brain works as a mechanistic input/output device, which simply couples sensory inputs to motor outputs that possibly cause a change in the inputs generating a new cycle. In the creative model, the brain represents the inputs on the basis of its spontaneous and intrinsic activity. This spontaneously active representation results in models of the world. These models might give rise to motor outputs (right lighter big arrow), which originates a feedback to the brain. (right lighter thin arrow). On the basis of this feedback, the brain can: 1. Change the way it processes sensory inputs (left thin arrow) or 2. Change the motor output (right darker big arrow), which might result in another feedback (right darker thin arrow) and eventually in new motor outputs, and so on. The crucial point is that the brain activity is spontaneous and does not depend on sensory inputs from the external.

external. The organism's expectation should not be assumed solely as an explicit, aware anticipation of possible events/stimuli but rather as a combination of several factors (including aware, unaware, cognitive, and emotional factors), which function as a bias or background for perception (i.e., for processing external inputs).

This account of evaluation resulting from the combination of expectations and perceptions is consistent with the view of the brain as more than a simple input/output device. In fact, since the 1980s, at least, advances in neuroscience have gone towards a shift from the view of the brain as a merely reactive input/output machine to its description as a spontaneous and intrinsically active organ (Changeux 1986, 2004). This results in a "creative model of the brain" (Farisco 2019; Figure 9.1).

While the view of the brain as a predictive, inferential machine is not new (Von Helmoltz 1867), the amount of empirical evidence converging in the creative model of the brain is innovative. As said, since the 1980s, neuroscience has gradually moved away from depicting the brain as a simple mechanistic input–output device towards a view that describes it as a complex, dynamic, and plastic organ that is spontaneously active. The traditional model of the brain can be called the reactive model: it depicts the brain as driven by inputs and producing outputs (i.e., as simply reactive to external stimuli). The new, creative model of the brain stresses that the brain itself is an active contributor (i.e., it actively interferes with and shapes the received stimuli in virtue of its spontaneous activity) (Changeux 1986; Edelman 1987; LeDoux 2002; Evers 2009; Laureys 2015; Raichle 2015; Singer 2015; Northoff 2018; Buzsáki 2019) (Figure 9.1).

A corollary of this model is that the brain is able to build predictions and models of the world that are useful in terms of interests, needs, and survival (Hohwy and Seth 2020). The inferential character of brain operations makes it possible to abstract away from actual stimuli, as illustrated by hallucinations or the inference of possible threats and opportunities coming from the environment. The creation of cultural and social categories and values also illustrates the capacity of the brain to "virtualize", that is, to abstract away from actual experience. This capacity for virtualization has been identified as one of the main hallmarks of awareness within a EU-funded interdisciplinary research project called Counterfactual Assessment and Valuation for Awareness Architecture (CAVAA), which aims to engineer aware AI systems (https://cavaa.eu/). Specifically, the capacity for virtualization derives from the fact that the brain is an intrinsically active and predictive organ, which explores the world projecting towards it the models (i.e., coherent neuronal patterns which underlie particular, consistent cognitive maps) that it spontaneously elaborates. The emotional salience of these models derives from their connection with the organism's needs and drives: this connection eventually makes the organism not indifferent to its own experiences, in the sense that "there is something it is like" for the organism to have those experiences.

This is the ground for the evolution of consciousness (i.e., subjective experience). In fact, emotions and the related capacity for evaluation "constitute [the organism's] motivation and volition, which increases its capacity to exert control over itself in interaction with its environment that it is motivated actively to explore" (Evers 2005).

9.3 The Role of Evaluation for Ethical Deliberation

The capacity of the human brain for evaluation is also the basis for the evolution of ethical deliberation, that is, the capacity for distinguishing between right and wrong. We may say that the basic mechanism underlying ethical evaluation is the same described previously as making possible to assess the valence of processed information for the organism's survival: on the basis of background priors, the brain elaborates models that are confronted with incoming actual information to be eventually confirmed or modified/adapted. Against this background, the brain assesses the valence of perceived information.

Very roughly, the process of ethical deliberation replicates the previous one, with additional factors playing a role in the decision-making process. In fact, ethical deliberation derives from the combination of individual and social priors, that is, from the background interests, needs, and goals of the agent and from the background interests, needs, and goals of the socio-cultural contexts where the agent has been educated and behaves. Importantly, these socio-cultural factors are not merely external to the agent but are "internalized" through a dynamic of cognitive and cerebral adaptations.

The epigenetic theory of neuronal development by selective stabilization of synapses (Changeux, Courrège et al. 1973) provides an explanation of the impact that the external environment has on the individual brain, with a number of effects at different levels, including ethical deliberation. Starting from the observation of a non-linear evolution between the complexity of the genome and that of the brain, which cannot be simply related to genome size, nor to number of genes, this theory proposes that the relation between brain and external environment (e.g., ecological, cultural, social, etc.) is critical for the development of the brain. The long postnatal period of brain maturation is the period when the brain is first subject to an explosive growth of neurons and synapses and then to a process of pruning that deeply shapes the brain. This process is led by genetic information interacting with external, contextual information. Both intrinsic and evoked brain activity can contribute to synapse selection through reward processes, which may be triggered by social (including ethical) values that eventually "leave a trace" in the brain architecture at both structural and functional levels. This scientific model has been explicitly translated into an ethical argument towards a pro-active ethical "care" of the developing brain (Evers and Changeux 2016).

Among other things, the epigenetic architecture of the brain affects the capacity for ethical deliberation, combining both intrinsic (e.g., emotional) and extrinsic (e.g., social) factors. In fact, we are born as essentially evaluative creatures but not with exactly the same evaluative or preferential tendencies (i.e., a tendency to develop a certain preference or a tendency to endorse a certain value) (Evers 2009). This raises the necessity of mediating between different values, which are socially and culturally shaped (Farisco 2024).

In short, evaluation (i.e., the capacity to assign positive and negative values to objects and experiences) plays a crucial role also in ethical deliberation, because it provides the subjects the tools for elaborating ideas about the "good" versus the "bad" and eventually choosing between what is good and what is bad. Importantly, the factors affecting evaluation and eventually ethical deliberation are not only internal but also external to the subject (i.e., socio-cultural factors); epigenetic dynamics guide the stabilization of some synapses and the pruning of others, eventually making possible the internalization of the effects of external factors. The actual technical feasibility of the last point in particular remains to be explored. In fact, even though there are recent computational studies on cultural evolution that show how AI agents can share information from each other and learn to perform tasks (arXiv:2206.05060) (Colas, Karch et al. 2022), as well as some recent attempts to inform robots with a capacity of interactive and open-ended learning (arXiv:2403.02514), at this stage AI mostly lacks an epigenetic development, at least of the same kind of that of humans. This may be a stumbling block in the development of artificial moral agents (i.e., artificial systems capable of ethical deliberation).

9.4 A Model for Artificial Evaluation

As argued previously, evaluation (i.e., the capacity for assigning positive and negative values to things) is a key evolutionary factor in biological organisms, and it also plays a crucial role for ethical deliberation, which is importantly shaped by both factors internal and external to the agent. We mentioned at the end of the last section that the (still) limited capacity of artificial systems for an epigenetic evolutionary adaptation may be a challenge for developing artificial moral agents. Notwithstanding this possible obstacle, there are technical attempts to advance in the direction of informing artificial systems with the capacity for evaluation and on that basis with the capacity for ethical deliberation as well.

For instance, some researchers in AI working in the previously mentioned project, CAVAA, are developing a computational model of virtualization (i.e., of internal representation or model of possible future scenarios and/or consequences of the system's actions or related events) which allows the system to identify which actions may have an impact on themselves or others that deserve specific consideration. Formalized within reinforcement learning (RL) theory (i.e., a machine learning paradigm that basically consists in training an algorithm to achieve desired results through a reward-based interaction with the environment) (Sutton and Barto 2018), this model is also conceived to replicate ethical reasoning, but it appears limited in its capacity to include factors external to the agent that, as seen previously, are ultimately key components of moral agency. Despite this limitation, a positive and potentially innovative feature of this attempt is the system's ability to plan multiple steps ahead, also through a counterfactual representation of anticipated actions' consequences, and to estimate uncertainty about them. Specifically, on the basis of this model, the system is able to assess multiple dimensions of these consequences: homeostatic (i.e., things that are important for the survival of the agent), epistemic (i.e., things that have the potential to bring new knowledge or information for the agent, eventually reducing uncertainty (entropy) in its internal representations/models of the world), and prospectively moral (i.e., the potential harm to other agents). Another important feature of this model is its ability to evaluate the relative importance of different actions' consequences, eventually balancing the expected value of the considered action in each dimension (homeostatic, epistemic, moral) and the level of uncertainty about the estimated value.

The underlying computational architecture relies on a robot reinforcement learning model (Dromnelle, Renaudo et al. 2023) which made it possible for the system to learn from two types of human feedback: evaluative feedback (e.g., correct/incorrect labelling of the system's actions) and demonstration (e.g., the robot sees the human acting and learns from the action's observed consequences).

On the basis of the process of virtualization summarized here, the developers of this model think that it promises to inform artificial systems with the capacity for ethical reasoning, which would eventually justify the attribution of moral agency to them. We think that this conclusion is not actually justified. First, "reasoning" is too anthropomorphic a term to be used in this context. In fact, the system computes rather than reasons (Marcus and Davis 2019; Santacaterina 2023). Second, the system does not really "make decisions" (for which background capacities like autonomy, will, and having preferences, among others, are needed); it rather computes relevant variables and produces outputs as specific functions. Third, even if we assumed that the previous model allows the system to make morally relevant decisions, this would not entail or even partly justify the conclusion that the system is capable of ethical reasoning (and therefore that it is a moral agent). In fact, in principle, it is possible to make morally relevant decision without realizing it, that is, without actually being morally responsible for that decision. As mentioned, present AI systems may be equipped with the capacity for evaluation (Dromnelle, Renaudo et al. 2023), which is eventually instrumental for achieving the goals they have been trained for, but it is not clear if it would really feel like something for AI to achieve a goal. On the contrary, it seems that current AI lacks any emotional stake in its interaction with the world so that it does not really make a difference for AI to function in one way rather than in another way. Also, AI's capacity for epigenetic (e.g., sociocultural) development is still very limited. Given the important role for ethical reasoning and moral agency played by the evaluative capacity of the human organism, mediated by internal features like emotions and external factors like culture, these limitations of actual AI are a potential stumbling block on the way to artificial moral agency.

Finally, the fundamental question of whether it is desirable to try to make AI a moral agent is worth being considered. It may be that to optimize the utility and efficiency of AI systems in particular contexts and for very sensible tasks (e.g., in schools for educating young children), the capacity for moral reasoning is crucial for maximizing the positive effects of using AI. Yet in principle this should not be taken from granted, for the very simple reason that the capacity for moral reasoning has had both beneficial and detrimental effects for humanity. We cannot exclude a priori that a system unable to perform moral reasoning is eventually also more beneficial from moral points of view than a moral subject.

9.5 Conclusion

Contemporary artificial systems are far from performing any functions that justify the ascriptions of mental terms, including the concept of "evaluation". The processing of data can lead to outcomes that may trigger anthropocentric reactions in some humans, such as making them say that they "collaborate" with the AI that "makes decisions", but in reality nothing like this really occurs. To compare, the car with automatic gear does not "evaluate" which gear to activate or "collaborate" with the driver (who simply uses the device). Presently, there seems to be no technical, empirical, or even theoretical foundation for assuming there is a capacity for evaluation in artificial systems, particularly for evaluation aimed at ethical deliberation.

Considering the emotional and cognitive complexity involved when humans make moral judgments and decisions, it seems highly unlikely that artificial systems will be even remotely close to performing this in the foreseeable future.

Considering the often heinous effects of the moral judgments and decisions that humans actually make and have made repeatedly through history (e.g. resulting in genocides of undesired human populations of varying identity; ecocides of natural ecosystems; or violent repression of social groups based on gender, race, ethnicity, or social class), it is by no means obvious that this capacity programmed into artificial systems by human beings would be less prone to such heinous effects. Caution in the attempts to do so is therefore called for.

Acknowledgements

This research has received funding from the project Counterfactual Assessment and Valuation for Awareness Architecture – CAVAA (European Commission, EIC 101071178).

References

Adolphs, R. and D. J. Anderson (2018). *The neuroscience of emotion: a new synthesis*. Princeton, Princeton University Press.

Arnold, M. B. (1960). *Emotion and personality*. New York, NY, Columbia University Press.

Bennett, M. S. (2023). *A brief history of intelligence: evolution, AI, and the five breakthroughs that made our brains*. New York, Mariner Books.

Birhane, A. (2021). "Algorithmic injustice: a relational ethics approach." *Patterns* 2(2): 100205.

Buzsáki, G. (2019). *The brain from inside out*. New York, NY, Oxford University Press.

Changeux, J.-P. (1986). *Neuronal man: the biology of mind*. New York, Oxford University Press.

Changeux, J.-P. (2004). *The physiology of truth: neuroscience and human knowledge*. Cambridge, MA, Belknap Press of Harvard University Press.

Changeux, P. Courrège and A. Danchin (1973). "A theory of the epigenesis of neuronal networks by selective stabilization of synapses." *Proceedings of the National Academy of Sciences of the United States of America* 70(10): 2974–2978.

Coeckelbergh, M. (2020). *AI ethics*. Cambridge, MA, The MIT Press.

Colas, C., T. Karch, C. Moulin-Frier and P.-Y. Oudeyer (2022). "Language and culture internalization for human-like autotelic AI." *Nature Machine Intelligence* 4(12): 1068–1076.

Damasio, A. R. (1994). *Descartes' error: emotion, reason, and the human brain*, Putnam.

Dehaene, S. and J. P. Changeux (1989). "A simple model of prefrontal cortex function in delayed-response tasks." *Journal of Cognitive Neuroscience* 1(3): 244–261.

Dignum, V. (2019). *Responsible artificial intelligence: how to develop and use AI in a responsible way*. Verlag, Springer.

Dreyfus, H. L. (1972). *What computers can't do: a critique of artificial reason*. New York, Harper & Row.

Dromnelle, R., E. Renaudo, M. Chetouani, P. Maragos, R. Chatila, B. Girard and M. Khamassi (2023). "Reducing computational cost during robot navigation and human–robot interaction with a human-inspired reinforcement learning architecture." *International Journal of Social Robotics* 15(8): 1297–1323.

Edelman, G. M. (1987). *Neural Darwinism: the theory of neuronal group selection*. New York, Basic Books.

Edelman, G. M. (1992). *Bright air, brilliant fire: on the matter of the mind*. New York, NY, BasicBooks.

EDPS (2020). Opinion on the European Commission's white paper on artificial intelligence – a European approach to excellence and trust (Opinion No. 4/2020).

Ekman, P. (1992). "An argument for basic emotions." *Cognition and Emotion* 6(3–4): 169–200.

Evers, K. (2005). "Neuroethics: a philosophical challenge." *American Journal of Bioethics* 5(2): 31–33; discussion W33–W34.

Evers, K. (2009). *Neuroetique. Quand la matière s'éveille*. Paris, Odile Jacob.

Evers, K. and J. P. Changeux (2016). "Proactive epigenesis and ethical innovation: a neuronal hypothesis for the genesis of ethical rules." *EMBO Reports* 17(10): 1361–1364.

Farisco, M. (2019). *Brain, consciousness and disorders of consciousness at the intersection of neuroscience and philosophy*. PhD, Uppsala University.

Farisco, M., Ed. (2024). *Neuroethics and cultural diversity*. London, Wiley-ISTE.

Farisco, M., G. Baldassarre, E. Cartoni, A. Leach, M. A. Petrovici, A. Rosemann, A. Salles, B. Stahl and S. J. van Albada (2024). "A method for the ethical analysis of brain-inspired AI." *Artificial Intelligence Review* 57(6): 133.

Farisco, M., K. Evers and A. Salles (2020). "Towards establishing criteria for the ethical analysis of artificial intelligence." *Science and Engineering Ethics* 26: 2413–2425.

Friedrich, A. B., J. Mason and J. R. Malone (2022). "Rethinking explainability: toward a postphenomenology of black-box artificial intelligence in medicine." *Ethics and Information Technology* 24(1): 8.

Haidt, J. (2003). *The moral emotions. Handbook of affective sciences*. New York, NY, Oxford University Press: 852–870.

Hohwy, J. and A. K. Seth (2020) "Predictive processing as a systematic basis for identifying the neural correlates of consciousness." *Philosophy and the Mind Sciences* 1(II).

Kaplan, A. and M. Haenlein (2019). "Siri, Siri, in my hand: Who's the fairest in the land? On the interpretations, illustrations, and implications of artificial intelligence." *Business Horizons* 62: 15–25.

Lane, R. D., L. Nadel and G. Ahern (2000). *Cognitive neuroscience of emotion*. New York, Oxford University Press.

Laureys, S. (2015). *Un si brillant cerveau. Les états limites de consciences*. Odile Jacob.

LeDoux, J. E. (1998). *The emotional brain: the mysterious underpinnings of emotional life*. New York, Simon & Schuster.

LeDoux, J. E. (2002). *Synaptic self: how our brains become who we are*. New York, Viking.

Marcus, G. and E. Davis (2019). *Rebooting AI: building artificial intelligence we can trust*. New York, Pantheon Books.

Misselhorn, C. (2022). "Artificial moral agents: conceptual issues and ethical controversy." *The Cambridge handbook of responsible artificial intelligence: interdisciplinary perspectives*. S. Voeneky, P. Kellmeyer, O. Mueller and W. Burgard. Cambridge, Cambridge University Press: 31–49.

Moll, J., R. Zahn, R. de Oliveira-Souza, F. Krueger and J. Grafman (2005). "The neural basis of human moral cognition." *Nature Reviews Neuroscience* 6(10): 799–809.

Nemitz, P. (2018). "Constitutional democracy and technology in the age of artificial intelligence." *Philosophical Transactions of the Royal Society A: Mathematical, Physical and Engineering Sciences* 376(2133): 20180089.

Nishant, R., M. Kennedy and J. Corbett (2020). "Artificial intelligence for sustainability: Challenges, opportunities, and a research agenda." *International Journal of Information Management* 53: 102104.

Northoff, G. (2018). *The spontaneous brain: from the mind-body to the world-brain problem*. Cambridge, MA, The MIT Press.

Parliament, E. (2020). *The ethics of artificial intelligence: issues and initiatives (No. PE 634.452)*. Bruxelles, EPRS | European Parliamentary Research Service.

Pessoa, L. (2009) "Cognition and emotion." *Scholarpedia* 4, 4567. https://doi.org/10.4249/scholarpedia.4567.

Raichle, M. E. (2015). "The restless brain: how intrinsic activity organizes brain function." *Philosophical Transactions of the Royal Society of London. Series B, Biological Sciences* 370(1668).

Rolls, E. T. (2005). *What are emotions, Why do we have emotions, and what is their computational basis in the brain? Who needs emotions? The brain meets the robot*. New York, NY, Oxford University Press: 117–146.

Santacaterina, M. (2023). *Adaptive resilience*. Hoboken, NJ, Wiley.

Singer, W. (2015). "The ongoing search for the neuronal correlate of consciousness." *Open MIND*. T Metzinger and J. Windt. Frankfurt am Main, MIND Group.

Stahl, B. C. (2021). "Artificial intelligence for a better future: an ecosystem perspective on the ethics of AI and emerging digital technologies." *SpringerBriefs in research and innovation governance*. Cham, Switzerland, Springer: 1, online resource.

Sutton, R. S. and A. G. Barto (2018). *Reinforcement learning: an introduction*. Cambridge, MA, The MIT Press.

Team, A. N. P. (2018). *The Toronto Declaration: Protecting the right to equality and non-discrimination in machine learning systems*. Toronto, Access Now.

Von Helmoltz, H. (1867). *Handbuch der Physiologischen Optik*. Leipzig, Leopold Voss.

Walton, N. and B. S. Nayak (2021). "Rethinking of Marxist perspectives on big data, artificial intelligence (AI) and capitalist economic development." *Technological Forecasting and Social Change* 166: 120576.

Whitby, B. (1991). "Ethical AI." *Artificial Intelligence Review* 5: 201–204.

Wiener, N. (1954). *The human use of human beings: cybernetics and society*. Boston, Houghton Mifflin.

Yeung, K. (2018). "Algorithmic regulation: a critical interrogation." *Regulation & Governance* 12(4): 505–523.

Zuboff, S. (2019). *The age of surveillance capitalism: the fight for a human future at the new frontier of power*. New York, PublicAffairs.

10

The Neuroscience of Humor

Shelia M. Kennison

10.1 Introduction

Humor occurs in all human societies and plays an integral role in most of our interactions with others (Jiang et al., 2019; Lu, 2023). Over the last fifty years, research has shed light on both psychological and neuroscientific aspects of humor. In this chapter, we review this research. First, the chapter summarizes the historical origins of the research and the view that humor-related behavior may have played an important role in the survival of humans and other species. Second, the chapter reviews the research on humor and the brain, including how humor appreciation changes due to changes in the brain, including a) during healthy aging, b) by brain injury due to accidents and stroke (i.e., disruption of blood flow in the brain), and c) by brain injury that occurs due to diseases that cause the destruction of brain tissue (e.g., Alzheimer's dementia, Parkinson's disease, Huntington's disease, among others). The section concludes with a discussion of the results of brain imaging studies, which identify multiple regions of the brain that are involved in the processing of humor. The last section focuses on research examining the effectiveness of humor therapies or interventions that promote mental and physical health. The chapter concludes with a description of studies that have tested the effectiveness of humor interventions in state-of-the-art randomized controlled clinical trials (RCTs).

10.2 Historical Perspectives on the Neuroscience of Humor

The scientific study of humor has a relatively short history, with an increase in studies occurring approximately fifty years ago (See Kennison, 2020; Martin and Ford, 2018 for review). Despite the decades of research on humor, no single theory has yet emerged to account for all humor-related behavior. The evolutionary psychological approach to humor is among the most prominent. The approach builds on Darwin's (1859) theory of natural selection (i.e., survival of the fittest), which explained how advantages of the physical features of organisms directly contributed to their survival. Organisms without those features did not survive and were only found in the fossil record.

DOI: 10.1201/9781032717791-10

Characteristics of organisms that are believed to provide a survival advantage are referred to as adaptive (Darwin, 1859). Using the evolutionary psychology approach, researchers analyze contemporary human behaviors in terms of how they may have provided a survival advantage in the distant past. For example, people may like to gossip (i.e., share information about other people in their groups) because in the distant past, knowing about others in our groups may have led to a survival advantage stemming from individuals being able to avoid conflict with others and/or to benefit from stronger social ties (Dunbar, 1996). Regarding the survival advantage of humor, evolutionary psychology has pointed out that humor-related behavior in humans is also likely to have played a role in the survival of humans in our distant past (Gruner, 1978).

Humor experiences involve some amount of positive emotion. The term used to describe the positive emotion experienced from humor is mirth (Amir et al., 2015). For example, when comprehending a joke, mirth occurs when one *gets the joke* during the comprehension of the punchline. Mirth may be followed by smiling or laughing. Understanding the neuroscience of humor begins with the understanding of human emotional responses. The remainder of this section reviews similarities in the emotional expression (e.g., smiling and laughing) in humans and other species.

10.2.1 Smiling across Species

Darwin (1872/2007) observed emotional expressions in animals and humans, concluding that emotions are innate and are the result of our biological endowment in place at birth. The implication of the conclusion was that emotional expressions would be universal, operating similarly in all humans across all human societies. A century later, the predictions would be supported by seminal research conducted by Ekman and Friesen (1971). The research documented the universality of six emotions: happiness, sadness, anger, fear, surprise, and disgust. Subsequently, other emotions have been added to this list, including amusement, relief, ecstasy, naches (a Yiddish term for feelings associated with pride a parent feels observing their child's achievements), and the feeling one gets when confronted with a challenge (fiero, Ekman and Cordaro, 2011). Panksepp and Watt (2011) added the feeling of lust to this list of universal human emotions. There is some debate as to which of these are 'basic' emotions and which are higher-order emotions (Panksepp and Watt, 2011) or homeostatic states (Panksepp, 2007). It suffices, for the purpose of this chapter, to acknowledge that such a controversy exists.

Gruner (1997) proposed that smiling and laughing in humans can be viewed as originating in aggressive behaviors. He described laughter as originating as a *roar of triumph*. Similarities in the facial expressions of humans and other mammal species exist. In humans, smiling involves the tensing of facial muscles that pull up the lips, making the teeth visible. Many other mammal species (e.g., lions, bears, wolves, as well as others) also pull up their lips during aggressive displays, revealing their sharp teeth (Davila-Ross and Palagi, 2022).

In recent decades, there has been increased interest in the emotional displays of non-human species (see van Heijst et al., 2023, for review). Waller and Dunbar (2005) found that during play, chimpanzees exhibit two smile-like expressions: a) a relaxed open-mouth display and b) a bared-teeth display without vocalization. They found that the animals used the bared teeth display to communicate affinity. In contrast, the relaxed open-mouth display occurred during play. Samson and Waller (2010) concluded that chimpanzee's bared-teeth displays are not aggressive. Numerous studies have now documented similar facial expressions in other primate species, including wild gorillas (Palagi et al., 2007; Waller and Cherry, 2012), bonobos (Demuru et al., 2015), macaques (Pellis et al., 2011), captive capuchin monkeys (Visalberghi et al., 2006), lemurs (Palagi et al., 2014), and other species (e.g., dogs, Maglieri et al., 2022).

Gruner's (1978, 1997) views that humor in humans has its roots in aggression are consistent with the observation that prior to the 20th century, humor in humans was unambiguously aggressive, usually involving the act of ridiculing or degrading others. The aggressive nature of humor was a core aspect of the superiority theory of humor advocated by the 17th century philosopher Thomas Hobbes. Hobbes (2008) proposed that using humor to ridicule others leads to the result of making the speaker feel better about themselves. Even in ancient Greece, Plato (428–348 BC) emphasized the common experience of delighting in the misfortune of others (Plato in Philebus, reprinted in Morreall, 1987). One can also find references to laughter in the Bible, which reflect negative functions of humor (e.g., mocking others) almost exclusively. Gruner (1997) conducted an analyses of a large sample of jokes and found that most involved an element of aggression in which someone is denigrated, even if in a playful manner.

10.2.2 Laughter-Like Vocalizations across Species

Humor may result in laughter (e.g., *ha ha ha*, Provine, 2017). In recent decades, numerous studies have examined laughter in humans and laugh-like vocalizations in other species (Winkler and Bryant, 2021). Research by Provine and Yong (1991) carried out an early rigorous study on human laughter. They documented the acoustic characteristics of laughter, showing that on average a bout of laughter could contain between 1 to 16 *ha* notes (e.g., *ha ha ha*). Later research further showed that there is variability in individuals' laughter in terms of the number of *ha* notes as well as the inclusion of other sounds during laughing (e.g., whistle-type sounds and snorts, Bachorowski et al., 2001).

Recent work has pointed out that some laughter is voluntary or forced, produced in social settings in which it may be perceived as socially unacceptable not to laugh (Kosilo et al., 2021). For example, we may be chatting with a higher-status person, such as a boss or teacher, who tells a joke. We may not find the joke strongly amusing, but we may feel social pressure to laugh. Forced laughter differs from involuntary laughter, which occurs during tickling or when we experience something strongly amusing. For the remainder of this chapter, we focus on the naturally occurring non-forced or involuntary forms of laughter.

Studies of vocalizations occurring during play by non-human primates have observed vocalizations similar to human laughter (Davila-Ross et al., 2009, 2011). In a study comparing laughter in human infants and young gorillas, chimpanzees, and bonobos when they were tickled, Davila-Ross et al. (2009) observed similarities in the acoustics of the vocalizations. In a later study, Davila-Ross et al. (2011) observed chimpanzees at play and noted the laughter-like vocalizations that occurred. They found that chimpanzees produced laughter spontaneously as well as in response to another chimpanzee's laughter. Their results suggested that laughter functioned to continue the chimpanzee's play with a partner. They concluded that human laughter and chimpanzee laughter have their origins as vocal signals of the desire to play.

A growing number of studies have documented laughter-like vocalizations in other mammal species (Burke et al., 2022; Knutson et al., 1998; Panksepp and Burgdorf, 2003). For example, laughter-like vocalizations are produced by rats when they are tickled by humans or play with other rats (Burke et al., 2022; Knutson et al., 1998; Panksepp and Burgdorf, 2003). These vocalizations are below the level of human hearing but can be recorded and analyzed later. Other rats are able to hear the sounds. Researchers have documented laughter-like vocalizations occurring during play in dolphins (Blomqvist et al., 2005) and domesticated dogs (Simonet et al., 2005). Winkler and Bryant (2021) review the literature on vocalizations during play in non-human species. They identified numerous published articles demonstrating distinct vocalizations during play by numerous other species.

Research on rat vocalizations during play also found that the propensity to make the vocalizations appeared to be influenced by genetics (Panksepp and Burgdorf, 2000). In the research, multiple generations of rats were studied. The researchers noted that rats that produced the most vocalizations during tickling had ancestors in the lab who were also robust vocalizers during play. Other rats in the lab who vocalized to a lesser degree descended from ancestors in the lab who were also only modest vocalizers. The researchers further noted that the frequencies of sound that rats initiated during play was also related to genetics. These results suggest that there may be a genetic component in the human propensity for laughing and possibly humor appreciation.

10.2.3 Play across Species

Humor-related behavior (e.g., smiling and laughing) in humans and similar behaviors in other species tend to occur during play (Pellegrini et al., 2007). Play has been defined as activities in which there is positive emotion and a mutual understanding among participants that the events involve pretending. For example, domesticated dogs display play fighting behaviors in which they will chase another dog, grab them loosely with their jaws, and force them to the ground without causing injury (Rossi et al., 2018). In chimpanzees, play sessions also involve pushing, pulling, wrestling, gentle biting, and vocalizations that signal play rather than real aggression (Cordoni et al., 2023). Human play behaviors begin in childhood and include play aggression as well as behaviors related to performing social roles (e.g., playing mother or father to a doll) or social scripts (e.g., having dinner or going to a restaurant) (Pellegrini et al., 2007).

There are multiple theories regarding why humans and other animals engage in play (Pellegrini et al., 2007). All suggest that play provides individuals with the opportunity to practice important behaviors and social roles that are relevant to survival in a safe context. By experiencing a wide array of play interactions, humans and animals are able to innovate behaviors that may be useful in future non-play interactions. Play behaviors across species can also be useful in serious situations to diffuse conflict (Bateson and Martin, 2013). When delivered deftly, humor in humans can change the tone of an interaction from serious to playful, can revise the perceived intention of an unwelcome comment (i.e., *I was just joking*), and can increase rapport with others in the interaction (Martin and Ford, 2018). When used in this way, a joke or other types of humor can serve as a verbal invitation to play.

10.3 Humor and the Brain

The roots of play and the humorous behaviors occurring during play may lie in our biology (Pellegrini et al., 2007). Typically developing human brains develop ready to engage in play and humor. Understanding the neuroscience of humor in its entirety requires an understanding of the brain's role in humor comprehension (e.g., understanding and appreciating humor) and humor production (e.g., creating humor). Our understanding of the role of different parts of the brain in controlling specific behaviors and abilities begins in the mid-1800s when researchers began to identify the functions of specific brain areas (Gazzaniga, 2014). Paul Broca and Carl Wernicke identified different locations in the left side of the brain, which when damaged led to impairments in using language. Since then, researchers have developed detailed maps of the brain, showing numerous locations that are involved in specific sensory and cognitive processes and also networks of multiple brain regions that together contribute to bringing about complex abilities (Siddiqi et al., 2022).

Among the studies examining how the brain brings about complex behavior, those focused on understanding humor are a minority. Nevertheless, in the last fifty years, a picture of how humor comprehension occurs in real time is emerging (Kennison, 2020). In this section, we will briefly review three lines of research on this topic: a) changes in humor processing that occurs in healthy aging due to the effects of aging on the brain, b) humor deficits in brain injuries and diseases that cause degeneration of the brain, and c) insights from brain imaging studies with healthy adults.

10.3.1 Changes in Humor Processing due to Healthy Aging

Healthy aging involves multiple declines in cognitive processing abilities, such as people having difficulty remembering names and other types of information as they get older; temporarily forgetting words when speaking (i.e., tip-of-the-tongue state, Wei et al., 2024); having more speech errors (e.g., mixing up names); and having problems understanding nonliteral language (i.e., figurative language), such as metaphors, idioms, irony, and sarcasm (Westbury and Titone, 2011). Irony and sarcasm can be sources of humor. Multiple studies have demonstrated that older adults do not perceive humor (i.e., get the joke) as well as younger adults (Shammi and Stuss, 2003; Uekermann et al., 2006). In the study by Shammi and Stuss (2003), participants viewed multiple cartoons and were asked to select the one that was funny. In a second task, they were asked to match cartoons and punchlines. Although older adults performed more poorly on these tasks than younger adults, when asked to rate the funniness of the items, their funniness ratings were higher than those of younger adults. Since then, other studies have shown that older adults experience difficulties comprehending humor, especially humor that requires one to consider the knowledge or thoughts of others (also called theory of mind processing, Bischetti et al., 2019, 2023).

Poor theory of mind processing in older adults can lead to social gaffes carried out by others to be perceived as less funny than they are by older adults. Stanley et al. (2014) showed young and older adults a series of video clips in which there were social gaffes. For example, a video depicted a work setting in which a supervisor has a negative reaction to an overweight female employee who says she would like to be a cheerleader. Understanding a social gaffe requires an understanding of what the point of view of each individual, which enables one to appreciate whether there is inconsistency in the person's behavior. The results of the study showed that young participants found gaffes funny, but older adults did not.

The reasons for age-related declines in humor perception appear to be linked directly to age-related changes in the brain, such as reductions in brain volumes (Fujita et al., 2023) and the weakening of connections between brain regions (Hughes et al., 2019). These changes lead to reductions in short-term memory ability, which is used to carry out mental processing, such as holding a conversation; understanding information that one is reading or hearing; and retrieving information from memory, such as specific words in speech (Bischetti et al., 2023). Age-related declines in humor perception are affected by age-related brain changes in one particular brain region – the right temporoparietal junction (TPJ) (Cherry et al., 2005), a region known to be involved in processing others' states of mind, thoughts, and intentions. In a recent study, Hughes et al. (2019) showed that declines in theory of mind processing in older adults was related to the weakening of connections between the right TPJ and other parts of the temporoparietal regions.

10.3.2 Humor Deficits due to Brain Injury or Degeneration

In the early 1900s, there were observations that damage to the brain can result in the inability to get humor in written or spoken materials despite the fact that the individual's comprehension of language was intact (Head, 1926). The observations were made by physicians treating patients who had experienced disruptions to blood flow in the brain (i.e., stroke). Survivors of stroke may experience a loss of ability. When language is disrupted, they are described as having aphasia. Head (1926) observed someone who lost the ability to get humor from a joke or cartoon but who had a typical ability to understand humor before the stroke.

Beginning in the 1970s, researchers examined humor deficits in individuals with brain injury, comparing their performance with matched control participants who were similar in terms of demographic variables (i.e., age, sex, and education) but who did not have a history of brain injury (see Kennison, 2020 for review). The majority of studies documenting disruptions in humor processing following brain damage have focused on survivors of stroke (e.g., Bihrle et al., 1986; Heath and Blonder, 2005; Gardner et al., 1975; Wapner et al., 1981) and traumatic brain injury (Braun et al., 1989). The finding that emerged from multiple studies showed that humor ability was impaired more following damage to the right hemisphere than the left hemisphere of the brain. For example, Bihrle et al. (1986) showed participants a sequence of three cartoon panels with one of two final panels. Each of the two final panels provided an unexpected conclusion to the sequence, but one was humorous and the other serious. When participants were asked to identify which final panel was funny, participants with right hemisphere damage performed worse than participants with damage to the left hemisphere.

Brain injury can also occur due to diseases that cause the degeneration of brain tissue (Garrett and Hough, 2021). Some of the diseases that have been found to affect humor processing include Alzheimer's dementia (Luzzi et al., 2020), frontotemporal dementia (Clark et al., 2016), Parkinson's disease (Thaler et al., 2012), Huntington's disease (Saldert et al., 2010), amyotrophic lateral sclerosis (ALS, Bambini et al., 2020), and long-term abuse of alcohol (Karabanowicz et al., 2021). A surprising observations is that these diseases are associated with damage to different regions of the brain. Of these diseases, most cause widespread damage to the brain. Alzheimer's dementia causes damage first to the temporal lobes of the brain and then other regions. Frontotemporal dementia causes damage to the frontal and temporal lobes of the brain. Parkinson's disease causes damage to the substantia nigra, a small region in the basal ganglia. Huntington's disease causes damage to the caudate nucleus in the basal ganglia of the brain. ALS causes damage to the neurons involved in movement (i.e., motor neurons), which are located in the spinal cord, brainstem, and cerebral cortex.

The implication of this research is that in healthy brains, all of these regions could be playing a role in the processing of humor. The notion that there are numerous regions in the brain participating in the complex behavior of humor may initially seem surprising, as prior research has identified single brain locations involved in complex abilities (i.e., vision, hearing, etc.). For example, localized brain regions have been shown to be for responsible for visual processing (i.e., occipital lobes), auditory processing (i.e., temporal lobes), and speech production (i.e., Broca's area), among other human abilities (See Garrett and Hough, 2021, for review). Nevertheless, the notion that there are numerous regions of the brain involved in humor processing is supported by the latest brain imaging studies. Increasingly, brain researchers view complex human abilities as brought about through brain networks rather than single brain regions.

One of the most compelling discoveries is that the regions of the brain involved in experiencing the mirth or positive emotion caused by humor are distinct from the regions involved in processing other aspects of humor (see Kennison, 2020 for review). Initial support for this conclusion came from observations of individuals experiencing gelastic epilepsy, a form of epilepsy in which individuals laugh uncontrollably during a seizure (Demir et al., 2022). Some individuals with gelastic epilepsy experience laughter with mirth, but others experience laughter without mirth. Cases of laughter without mirth suggested that within the brain, the regions controlling mirth must be distinct from those controlling laughter alone.

There have now been studies utilizing brain stimulation techniques confirming the possibility of distinct brain regions for laugher with mirth and laughter without mirth (Caruana et al., 2015, 2016). During brain surgeries, surgeons can apply electricity directly to brain regions using a probe and then observe the effects of stimulating different brain regions brings about laughter with mirth and laughter without mirth. The procedure is referred to as electrical brain stimulation (EBS). In a study of patients with epilepsy, Caruana et al. (2015) investigated responses following the stimulation of the pregenual anterior cingulate; some patients produced laughter with mirth, and some patients produced laugher without mirth. In a follow-up study, Caruana et al. (2016) reported a study involving four patients with epilepsy who smiled and laughed without mirth following stimulation of the left and right frontal operculum, a region located below the inferior frontal gyrus (also referred to as Broca's area).

Using methods that enable the visualization of white matter connections in the brain, Gerbella et al. (2021) tested healthy participants and found that there are two distinct networks of brain areas involved in laughter production, one that is involved in producing laughter occurring with mirth and one involved in producing laughter without an emotional response, such as what occurs during forced laughter. The former network involved the ventral temporal pole, pregenual anterior cingulate, and ventral striatum/nucleus accumbens. The latter network involved the primary motor cortex and the frontal operculum. The two networks were connected in the pre-supplementary motor area.

10.3.3 Brain Imaging Studies of Humor

Neuroscientists have begun to investigate the regions of the brain involved in appreciating and comprehending humor (see Kennison, 2020 for review). There have now been multiple studies using the state-of-the-art functional magnetic resonance imaging (fMRI), which provides detailed images of changes occurring in the brain, to observe brain activity during humor comprehension. The consensus emerging from multiple fMRI studies of humor comprehension disconfirms the view that a single brain region is responsible for humor processing and supports a more complicated reality of numerous brain regions working together as a circuit.

Multiple fMRI studies have investigated the comprehension of humor that involves an incongruity (see Chan, 2024, for review). The processing of the cartoon caption or joke punchline was of primary interest. Researchers found that the comprehension of humor involved multiple stages of processing. Three stages of processing have been identified in multiple studies. The first stage involves the detection of incongruity (i.e., a violation of an expectation) and is associated with activity in the medial frontal gyrus and the medial temporal gyrus. The second stage involves the resolution of the incongruity (i.e., a humorous interpretation is found) and is associated with activity in five regions. Two of the five regions are known language areas (i.e., the inferior frontal gyrus or Broca's area and posterior superior temporal gyrus or Wernicke's area), and a third is known to be involved in the processing of others' minds (i.e., the temporoparietal junction). The remaining areas involved in the resolution of incongruity have been involved in memory processing (i.e., superior frontal gyrus) and processing emotions of others (i.e., inferior parietal lobule). In the third stage, participants elaborate on the humor to experience the pleasure of the humor (i.e., positive emotion), associated with activity in eleven areas: ten involved in the emotional response (i.e., the amygdala, temporal pole, superior temporal sulcus, anterior insula, ventral striatum, orbitofrontal cortex, ventral tegmental area, substantia nigra, ventromedial prefrontal cortex, and ventral anterior cingulate cortex) and one involved in the movement (i.e., supplementary motor area).

In a recent study by Chang et al. (2023), four stages are proposed. The last three stages are the same as those described by Chan (2024). Chang et al. (2023) suggested that the first stage is one in which the initial context of the humor or setup is processed. This stage involves the development of an expectation, which is later violated during the incongruity detection stage. They observed that the processing in the first stage was associated with activity in the medial frontal gyrus and the posterior insula.

In sum, the processing of humor involves many regions of the brain working together. We can imagine that the coordination of processing occurs rapidly without our ability to introspect about how we are completing the task. In the earliest stages of processing humor, we may not be aware that we will soon find the conclusion funny. We are using the parts of our brain that carry out comprehension of serious material. When we detect an incongruity and then resolve it to discover a humorous possibility, then the areas of our brain involved in experiencing pleasure (i.e., the reward center) become involved. If the humor is strong enough to cause smiling, laughter, and even movement of the body, such as hand movements to slap a knee or to sway, bending the knees, then areas of the brain controlling body movements also become involved.

10.4 Harnessing Humor for Health

Decades of research have documented the fact that experiencing humor leads to physiological changes in the body, such as increases in heart and lung function, reductions in the stress hormones and pain, and changes in metabolism that lead to reductions in blood glucose levels (See Oliveira et al., 2023 for review). Activity in regions of the brain involved in humor processing is likely to be directly linked to some or all of the physiological changes that have been observed in the body. Because many of the changes in the body that occur during the processing of humor are beneficial, researchers have begun to develop humor interventions for the purpose of improving mental and/or physical health. An increasing number of studies have tested the benefits of humor on health using the state-of-the-art techniques, including RCTs in which groups of individuals are randomly assigned to receive a humor intervention and another group (i.e., the control group) receives a similar experience that does not involve humor. This final section of this chapter reviews the research demonstrating the benefits of humor on health. Future research will be needed to detail how the health benefits of experiencing humor stem from activity in specific brain regions.

10.4.1 Early Anecdotal Evidence of the Healing Power of Humor

The power of humor has been discussed in specific cases in which people have endured extreme stress. Survivors of imprisonment during the Vietnam War reported using humor to cope with the difficult conditions experienced as a prisoner of war (POW) (Henman, 2001). In the documentary film *The Last Laugh*, some survivors of the Holocaust remarked about their use of humor to cope with the dire conditions in concentration camps and also coping with memories of their trauma late in life (Pearlstein, 2016).

In the 1970s, Norman Cousins, a well-known writer and editor, wrote about his personal experiences using humor to cope with severe pain caused by ankylosing spondylitis, a form of arthritis affecting the spine (Cousins, 1979). The condition worsened over time. His doctors conveyed to him that the chances of recovering from the condition were small. He worked diligently, researching possible treatments to improve his quality of life. He also relied on humorous movies, TV shows, and books for some relief. Over time, he recorded his daily consumption of humor and his pain. He reported that merely ten minutes of hearty laughter produced about two hours of pain reduction, allowing him to sleep. His first-person account of the benefits of humor increased interest in the topic. In the decades since, there have been thousands of scientific studies on the mental and physical health benefits of humor.

10.4.2 Empirical Evidence of the Healing Power of Humor

Since the 1980s, there have been numerous studies examining the effects of humor on physical processes of the body that are known to be related to health (Oliveira et al., 2023). Health researchers have also reported that coping with humor can be beneficial for patients facing a range of serious health conditions, including cancer (Kim et al., 2015; Shi et al., 2024), cardiovascular disease (Lockwood and Yoshimura, 2014), and chronic kidney disease (Sousa et al., 2019). Most of these studies relied on research methodologies that did not enable the researchers to determine with certainty that experiencing humor caused improvements in health (e.g., case studies, qualitative analyses, and/or correlational study designs).

Experiencing humor activates the sympathetic nervous system (i.e., the body's fight-or-flight system), increasing one's heart rate and breathing rate (Lackner et al., 2014). The act of laughing during a humor experience involves deep inhalation, which increases the oxygen levels of the blood, and the rapid exhalation during the production of the laugh (i.e., *ha ha ha*). The muscles in the torso contract rapidly during the vocalization of laughter. When the humor is especially strong, one can experience a weakening in the muscles in the legs. The expression *weak in the knees* describes this effect of humor on the body.

Experiencing humor also actives the hypothalamic–pituitary–adrenocortical (HPA) axis, triggering physiological processes that lead to the lowering of stress hormones (e.g., cortisol, Meier et al., 2021) and stimulation of the immune system (Bennett et al., 2003; Hayashi et al., 2007). Experiencing humor can also lead to the lowering of blood glucose and pain. The reduction of pain following a strong humor experience may come about due to the activation of opioid receptors in the brain.

Although numerous studies have been conducted thus far, there are likely more health effects of humor that have not yet been identified. In the next section, we examine the latest studies testing the effects of humor using the state-of-the-art design used in medical research, the RCT.

10.4.3 Evidence from Randomized Controlled Trials

RCTs are routinely used in medicine and other disciplines to determine whether a new medication, new treatment, or new experimental manipulation has better outcomes when compared with existing medications, treatments, placebo control conditions, or other control conditions (Houle, 2015). A key feature of RCTs is that at the beginning of the study, participants are randomly assigned to conditions. Randomization reduces bias from being introduced into the study due to how participants get assigned to groups (Steeger et al., 2021). When using randomization, larger samples are less vulnerable to bias than small samples.

There are now numerous studies using RCTs to test the effects of humor on a self-reported mental and physical health as well as on objective measures of health (e.g., Cai et al., 2014; Harputlu et al., 2023; Hirosaki et al., 2023; Kim et al., 2015; Ko and Park, 2023; Kramer and Leitao, 2023; Rudnick et al., 2014; Sarink and García-Montes, 2023; Van der Wal and Kok, 2019; Wellenzohn et al., 2016; Zhao et al., 2019). For example, research using the RCT design has examined the effectiveness of humor referred to as the 'three funny things' intervention (Wellenzohn et al., 2016). In the intervention, participants were asked to briefly write each day for seven days either about: a) three funny things, which occurred in the present or the past or might occur in the future or b) a non-humorous topic (i.e., childhood memories). Participants' levels of happiness were higher and levels of depression were lower after the humor intervention. These changes were not observed for the participants in the control condition.

In a study of women receiving treatments for breast cancer, Kim et al. (2015) investigated the effect of humor on the health outcomes. Participants were randomly assigned to receive four sessions of humor or four sessions of a control intervention that did not involve humor. Participants in the humor condition showed reduced depression, anxiety, and stress, while those in the control condition did not.

In a recent study, Kramer and Leitao (2023) examined the relationship between spontaneous laughter and levels of the stress hormone cortisol across numerous studies, including both RCTs as well as experiments comparing different groups of participants who were not randomly assigned to conditions. Cortisol is a hormone that is produced during stress. They found that across studies, cortisol was lower when there was more frequent occasions of spontaneous laughter.

In a study that focused on participants with Type 2 diabetes, Hirosaki et al. (2023) tested the effectiveness of a humor intervention (i.e., laughter yoga) on multiple health-related outcomes, including hemoglobin A1c, which is a measure of the average blood glucose over the past three months. Forty-two participants were randomly assigned to the humor intervention or control (i.e., standard therapy and spending time as they usually would), which occurred over a twelve-week period. The results showed that those in the humor intervention group had improved A1c levels, better mood, and longer sleep than those in the control group.

In a study in which prior studies examining the effects of humor interventions were analyzed, Van der Wal and Kok (2019) concluded that the positive effects of laughter were larger when participants produced forced laughter (i.e., purposeful non-humorous laughter) as compared to when participants produced laughter while experiencing humor. Forced laughter is an integral part of laughing yoga (Kuru Alici and Arikan Dönmez, 2020), a form of yoga that involves meeting with others and producing laughter on cue as directed by a leader. Van der Wal and Kok's (2019) study is compelling because it suggests that people may be able to experience many of the positive effects of humor without actually experiencing the pleasurable experience of mirth, which is triggered spontaneously during humor processing.

Given the increasing number of well-designed and controlled studies showing that humor can cause health benefits, there is the promise that more healthcare professionals and individuals will find ways to incorporate humor interventions into daily life. It remains unknown how much of the benefit of humor stems from benefits derived from the physical activity of laughing and how much is derived from experiencing positive emotion or mirth that is the result of experiencing a humorous event. In terms of ease of implementation of humor interventions, institutional settings may provide the best opportunity to impact the most people in the shortest amount of time. Hospitals, mental health clinics, nursing homes, and assisted living settings are locations where humor interventions could be carried out with minimal cost.

References

Amir O, Biederman I, Wang Z, Xu X (2015). Ha Ha! Versus Aha! A direct comparison of humor to nonhumorous insight for determining the neural correlates of mirth. *Cereb. Cortex 25*: 1405–1413.

Bachorowski JA, Smoski MJ, Owren MJ (2001). The acoustic features of human laughter. *J. Acoust. Soc. Am. 110*: 1581–1597.

Bambini V, Bischetti L, Bonomi CG, Arcara G, Lecce S, Ceroni M (2020). Beyond the motor account of amyotrophic lateral sclerosis: Verbal humour and its relationship with the cognitive and pragmatic profile. *Int. J. Lang. Comm. Disord. 55*: 751–764.

Bateson PPG, Martin P (2013). *Play, playfulness, creativity and innovation*. Cambridge: Cambridge University Press.

Bennett MP, Zeller JM, Rosenberg L, McCann J (2003). The effect of mirthful laughter on stress and natural killer cell activity. *Altern. Ther. Health Med. 9*: 38–45.

Bihrle AM, Brownell HH, Powelson JA, Gardner H (1986). Comprehension of humorous and non-humorous materials by left and right brain-damaged patients. *Brain Cogn. 5*: 399–411.

Bischetti L, Ceccato I, Lecce S, Cavallini E, Bambini V (2019). Pragmatics and theory of mind in older adults' humor comprehension. *Curr. Psychol. 42:* 16191–16207.

Bischetti L, Ceccato I, Lecce S, Cavallini E, Bambini V (2023). Pragmatics and theory of mind in older adults' humor comprehension. *Curr. Psychol. 42*: 16191–16207.

Blomqvist C, Mello I, Amundin M (2005). An acoustic play-fight signal in bottlenose dolphins (*Tursiops truncatus*) in human care. *Aquat. Mamm. 31*: 187–194.

Braun CM, Lussier F, Baribeau JM, Ethier M (1989). Does severe traumatic closed head injury impair sense of humour? *Brain Inj. 3*: 345–354.

Burke CJ, Pellis SM, Achterberg EJM (2022). Who's laughing? Play, tickling and ultrasonic vocalizations in rats. *Philos. Trans. R. Soc. B 377:* 20210184.

Cai C, Yu L, Rong L, Zhong H (2014). Effectiveness of humor intervention for patients with schizophrenia: A randomized controlled trial. *J. Psychiatr. Res. 59*: 174–178.

Caruana F, Avanzini P, Gozzo F, Francione S, Cardinale F, Rizzolatti G (2015). Mirth and laughter elicited by electrical stimulation of the human anterior cingulate cortex. *Cortex 71*: 323–331.

Caruana F, Gozzo F, Pelliccia V, Cossu M, Avanzini P (2016). Smile and laughter elicited by electrical stimulation of the frontal operculum. *Neuropsychologia 89*: 364–370.

Chan YC (2024). The neuroscience of humor. In T. Ford, W. Chłopicki, G. Kuipers (Eds) *De Gruyter handbook of humor studies* (pp. 65–84). Degruyter.

Chang CY, Chan YC, Chen HC (2023). Verification of the four-stage model of humor processing: Evidence from an fMRI study by three-element verbal jokes. *Brain Sci. 13*: 417.

Cherry BJ, Adamson M, Duclos A, Hellige JB (2005). Aging and individual variation in interhemispheric collaboration and hemispheric asymmetry. *Aging Neuropsychol. Cogn. 12*: 316–339.

Clark CN, Nicholas JM, Gordon E, Golden HL, Cohen MH, Woodward FJ, Macpherson K, Slattery C, Mummery C, Schott J, Rohrer JD, Warren JD (2016). Altered sense of humor in dementia. *J. Alzheimers Dis. 49*: 111–119.

Cordoni G, Ciarcelluti G, Pasqualotto A, Perri A, Bissiato V, Norscia I (2023). Is it for real? Structural differences between play and real fighting in adult chimpanzees (*Pan troglodytes*). *Am. J. Primatol. 85:* e23537.

Cousins N (1979). *Anatomy of an illness*. New York: W. W. Norton & Company.

Darwin C (1859). *On the origin of species by means of natural selection*. London: J. Murray Publishing.

Darwin C (1872/2007). *The expression of the emotions in man and animals*. London: Filiquarian.

Davila-Ross M, Allcock B, Thomas C, Bard KA (2011). Aping expressions? Chimpanzees produce distinct laugh types when responding to the laughter of others. *Emot. 11*: 1013.

Davila-Ross M, Owren MJ, Zimmermann E (2009). Reconstructing the evolution of laughter in great apes and humans. *Curr. Biol. 19*: 1106–1111.

Davila-Ross M, Palagi E (2022). Laughter, play faces and mimicry in animals: Evolution and social functions. *Philos. Trans. R. Soc. B:* 20210177.

Demir AB, Öz BY, Yıldız MO, Türk BG, Tanrıverdi T, Bekar A, Yeni N, Bora İ (2022). A clinical evaluation of gelastic and dacrystic seizures: A multicenter study. *Arq. Neuro-Psiquiatr. 80*: 1204–1212.

Demuru E, Ferrari PF, Palagi E (2015). Emotionality and intentionality in bonobo playful communication. *Anim. Cogn. 18*: 333–344.

Dunbar RIM (1996). *Grooming, gossip, and the evolution of language*. Cambridge: Harvard University Press.

Ekman P, Cordaro D (2011). What is meant by calling emotions basic. *Emot. Rev. 3*(4): 364–370.

Ekman P, Friesen WV (1971). Constants across cultures in the face and emotion. *J. Pers. Soc. Psychol. 17*: 124–129.

Fujita S, Mori S, Onda K, Hanaoka S, Nomura Y, Nakao T, Yoshikawa T, Takao H, Hayashi N., Abe O (2023). Characterization of brain volume changes in aging individuals with normal cognition using serial magnetic resonance imaging. *JAMA Netw. Open 6*: e2318153.

Gardner H, Ling P, Flamm L, Silverman J (1975). Comprehension and appreciation of humour in brain-damaged patients. *Brain 98:* 399–412.

Garrett B, Hough G (2021). *Brain & behavior: An introduction to behavioral neuroscience* (6th edition). Los Angeles: Sage Publications.

Gazzaniga MS (2014). *Handbook of cognitive neuroscience*. Springer.

Gerbella M, Pinardi C, Di Cesare G, Rizzolatti G, Caruana F (2021). Two neural networks for laughter: a tractography study. *Cereb. Cortex 31*(2): 899–916.

Gruner C (1997). *The game of humor: A comprehensive theory of why we laugh.* New York: Transaction.

Gruner CR (1978). *Understanding laughter: The workings of wit & humor.* New York: Rowman & Littlefield.

Harputlu D, Öztürk FÖ, Aydın D, Akyol C, Tezel A (2023). Effect of laughter yoga on sleep and quality of life in individuals with fecal ostomies: A randomized controlled trial. *J. Wound Ostomy Cont. 50*: 307–312.

Hayashi T, Tsujii S, Iburi T, Tamanaha T, Yamagami K, Ishibashi R, Hori M, Sakamoto S, Ishii M, Murakami K (2007). Laughter up-regulates the genes related to NK cell activity in diabetes. *Biomed. Res. 28*: 281–285.

Head H (1926). *Aphasia and kindred disorders of speech*. Cambridge: Cambridge University Press.

Heath RL, Blonder LX (2005). Spontaneous humor among right hemisphere stroke survivors. *Brain Lang. 93*: 267–276.

Henman LD (2001). Humor as a coping mechanism: Lessons from POWs. *Humor 14*: 83–94.

Hirosaki M, Ohira T, Wu Y, Eguchi E, Shirai K, Imano H, Funakubo N, Nishizawa H, Katakami N, Shimomura L, Iso H (2023). Laughter yoga as an enjoyable therapeutic approach for glycemic control in individuals with type 2 diabetes: A randomized controlled trial. *Front. Endocrinol. 14*: 1148468.

Hobbes T (2008). *Leviathan* (JCA Gaskin, Ed.). Oxford University Press.

Houle S (2015). An introduction to the fundamentals of randomized controlled trials in pharmacy research. *Can. J. Hosp. Pharm. 68:* 28–32.

Hughes C, Cassidy BS, Faskowitz J, Avena-Koenigsberger A, Sporns O, Krendl AC (2019). Age differences in specific neural connections within the Default Mode Network underlie theory of mind. *NeuroImage 191*: 269–277.

Jiang T, Li H, Hou Y (2019). Cultural differences in humor perception, usage, and implications. *Front. Psychol. 10*: 123.

Karabanowicz E, Tyburski E, Karasiewicz K, Bober A, Sagan L, Mak M, Radziwiłłowicz W (2021). Higher-order language dysfunctions in individuals with alcohol use disorder. *Journal of Clin. Med. 10*: 4199.

Kennison SM (2020). *Cognitive neuroscience of humor.* Washington, DC: American Psychological Association.

Kim SH, Kook JR, Kwon M, Son MH, Ahn SD, Kim YH (2015). The effects of laughter therapy on mood state and self-esteem in cancer patients undergoing radiation therapy: A randomized controlled trial. *J. Altern. Complement. Med. 21*: 217–222.

Ko Y, Park S (2023). A pilot randomized controlled trial of distance laughter therapy for mothers' level of depression, anxiety, and parental stress during the COVID-19 pandemic. *PLoS One 18*: e0288246.

Kosilo M, Costa M, Nuttall HE, Ferreira H, Scott S, Menéres S, Pestana J, Jerónimo R, Prata D (2021). The neural basis of authenticity recognition in laughter and crying. *Sci. Rep. 11*: 23750.

Knutson B, Burgdorf J, Panksepp J (1998). Anticipation of play elicits high-frequency ultrasonic vocalizations in young rats. *J. Comp. Psychol. 112*: 65–73.

Kramer CK, Leitao CB (2023). Laughter as medicine: A systematic review and meta-analysis of interventional studies evaluating the impact of spontaneous laughter on cortisol levels. *PLoS One 18*: e0286260.

Kuru Alici N, Arikan Dönmez A (2020-11-01). A systematic review of the effect of laughter yoga on physical function and psychosocial outcomes in older adults. *Complement. Ther. Clin. Pract. 41*: 101252.

Lackner HK, Weiss EM, Hinghofer-Szalkay H, Papousek I (2014). Cardiovascular effects of acute positive emotional arousal. *Appl. Psychophysiol. Biofeedback 39*: 9–18.

Lockwood NL, Yoshimura SM (2014). The heart of the matter: The effects of humor on well-being during recovery from cardiovascular disease. *Health Comm. 29*: 410–420.

Lu JG (2023). Cultural differences in humor: A systematic review and critique. *Curr. Opin. Psychol. 53*: 101690.

Luzzi S, Baldinelli S, Ranaldi V, Fiori C, Plutino A, Fringuelli FM, Silvestrini M, Baggio, G, Reverberi C (2020). The neural bases of discourse semantic and pragmatic deficits in patients with frontotemporal dementia and Alzheimer's disease. *Cortex 128*: 174–191.

Maglieri V, Zanoli A, Mastrandrea F, Palagi E (2022). The relaxed open mouth is a true signal in dogs: Demonstrating Tinbergen's ritualization process. *Anim. Behav. 188*: 65–74.

Martin RA, Ford T (2018). *The psychology of humor: An integrative approach*. New York: Academic Press.

Meier M, Wirz L, Dickinson P, Pruessner JC (2021). Laughter yoga reduces the cortisol response to acute stress in healthy individuals. *Stress 24*: 44–52.

Morreall J (1987). *The philosophy of laughter and humor.* Albany, NY: State University of New York Press.

Oliveira R, Arriaga P, Barreiros J (2023). The role of humor in social, psychological, and physical well-being. *Humor 36*: 487–509.

Palagi E, Antonacci D, Cordoni G (2007). Fine-tuning of social play in juvenile lowland gorillas (gorilla gorilla gorilla). *Dev. Psychobiol. 49*: 433–445.

Palagi E, Norscia I, Spada G (2014). Relaxed open mouth as a playful signal in wild ring-tailed lemurs. *Am. J. Primatol. 76*: 1074–1083.

Panksepp J (2007). Criteria for basic emotions: Is DISGUST a primary "emotion"? *Cogn. Emot. 21*(8): 1819–1828.

Panksepp J, Burgdorf J (2000). 50-kHz chirping (laughter?) in response to conditioned and unconditioned tickle-induced reward in rats: Effects of social housing and genetic variables. *Behav. Brain Res. 115*: 25–38.

Panksepp J, Burgdorf J (2003). Laughing rats and the evolutionary antecedents of human joy? *Physiol. Behav. 79*: 533–547.

Panksepp J, Watt D (2011). What is basic about basic emotions? Lasting lessons from affective neuroscience. *Emot. Rev. 3*(4): 387–396.

Pearlstein F (2016). *The last laugh* [Film]. Tangerine Entertainment.

Pellegrini AD, Dupuis D, Smith PK (2007). Play in evolution and development. *Dev. Rev. 27*: 261–276.

Pellis SM, Pellis VC, Reinhart CJ, Thierry B (2011). The use of the bared-teeth display during play fighting in Tonkean macaques (*Macaca tonkeana*): Sometimes it is all about oneself. *J. Comp. Psychol. 125*: 393.

Provine RR (2017). Laughter as an approach to vocal evolution: The bipedal theory. *Psychon. Bull. 24*: 238–244.

Provine RR, Yong YL (1991). Laughter: A stereotyped human vocalization. *Ethol. 89*: 115–124.

Rossi A, Parada FJ, Stewart R, Barwell C, Demas G, Allen C (2018). Hormonal correlates of exploratory and play-soliciting behavior in domestic dogs. *Front. Psychol. 9*: 1559.

Rudnick A, Kohn PM, Edwards KR, Podnar D, Caird S, Martin R (2014). Humour-related interventions for people with mental illness: A randomized controlled pilot study. *Community Ment. Health J. 50*: 737–42.

Saldert C, Eriksson E, Petersson K, Hartelius L (2010). Interaction in conversation in Huntington's disease: An activity-based analysis and the conversation partner's view of change. *J. Interact. Res. Commun. Disord. 1*: 169–197.

Samson AM, Waller BM (2010). Not growling but smiling: New interpretations of the bared-teeth motif in the pre-Columbian Caribbean. *Curr. Anthropol. 51*: 425–433.

Sarink FS, García-Montes JM (2023). Humor interventions in psychotherapy and their effect on levels of depression and anxiety in adult clients, a systematic review. *Front. Psychiatry 13*: 1049476.

Shammi P, Stuss DT (2003). The effects of normal aging on humor appreciation. *J. Int. Neuropsychol. Soc. 9*, 855–863.

Shi H, Wu Y, Wang L, Zhou X, Li F. (2024). Effects of laughter therapy on improving negative emotions associated with cancer: A systematic review and meta-analysis. *Oncol. 102*: 343–353.

Siddiqi SH, Kording KP, Parvizi J, Fox MD (2022). Causal mapping of human brain function. *Nat. Rev. Neurosci. 23*: 361–375.

Simonet P, Versteeg D, Storie D (2005). Dog-laughter: Recorded playback reduces stress related behavior in shelter dogs. *Seventh International Conference on Environmental Enrichment*. New York: Columbia University.

Sousa LM, Antunes AV, Marques-Vieira CM, Silva PC, Valentim OM de S, José HM (2019). Subjective wellbeing, sense of humor and psychological health in hemodialysis patients. *Enferm. Nefrol. 22*: 34–41.

Stanley JT, Lohani M, Isaacowitz DM (2014). Age-related differences in judgments of inappropriate behavior are related to humor style preferences. *Psychol. Aging 29*(3): 528.

Steeger CM, Buckley PR, Pampel FC, Gust CJ, Hill KG (2021). Common methodological problems in randomized controlled trials of preventive interventions. *Prev. Sci. 22*: 1159–1172.

Thaler A, Posen J, Giladi N, Manor Y, Mayanz C, Mirelman A, Gurevich T (2012). Appreciation of humor is decreased among patients with Parkinson's disease. *Parkinsonism Relat. D. 18*: 144–148.

Uekermann J, Channon S, Daum I (2006). Humor processing, mentalizing, and executive function in normal aging. *J. Int. Neuropsychol. Soc. 12*: 184–191.

van Heijst K, Kret ME, Ploeger A (2023). Basic emotions or constructed emotions: Insights from taking an evolutionary perspective. *Perspect. Psychol. Sc.* 17456916231205186.

Van der Wal CN, Kok RN (2019). Laughter-inducing therapies: Systematic review and meta-analysis. *Soc. Sci. Med. 232:* 473–488.

Visalberghi E, Valenzano DR, Preuschoft S (2006). Facial displays in Cebus apella. *Int. J. Primatol. 27*: 1689–1707.

Waller BM, Cherry L (2012). Facilitating play through communication: Significance of teeth exposure in the gorilla play face. *Am. J. Primatol. 74:* 157–164.

Waller BM, Dunbar RI (2005). Differential behavioural effects of silent bared teeth display and relaxed open mouth display in chimpanzees (*Pan troglodytes*). *Ethology 111*(2): 129–142.

Wapner W, Hamby S, Gardner H (1981). The role of the right hemisphere in the apprehension of complex linguistic materials. *Brain Lang. 14*: 15–33.

Wei HT, Kulzhabayeva D, Erceg L, Robin J, Hu YZ, Chignell M, Meltzer JA (2024). Cognitive components of aging-related increase in word-finding difficulty. *Aging Neuropsychol. Cogn*. 1–32.

Wellenzohn S, Proyer RT, Ruch W (2016). Humor-based online positive psychology interventions: A randomized placebo-controlled long-term trial. *J. Posit. Psychol. 11*: 584–594.

Westbury C, Titone D (2011). Idiom literality judgments in younger and older adults: Age-related effects in resolving semantic interference. *Psychol. Aging 26*(2): 467.

Winkler SL, Bryant GA (2021). Play vocalisations and human laughter: A comparative review. *Bioacoustics 30*: 499–526.

Zhao J, Yin H, Zhang G, Li G, Shang B, Wang C, Chen L (2019). A meta-analysis of randomized controlled trials of laughter and humour interventions on depression, anxiety and sleep quality in adults. *J. Adv. Nurs. 75*: 2435–2448.

11

The Neuroscientific Study of Religious and Spiritual Phenomena

Andrew B. Newberg

11.1 Introduction

The growing realm of research delving into religious and spiritual phenomena sparks various viewpoints regarding its validity, significance, and relevance. It is a complex and multidisciplinary field, sometimes referred to as *neurotheology*. Evaluating this field entails identifying current methodological challenges and devising strategies to address them, aiming to enhance the robustness of future inquiries and integrate this research into mainstream discourse. Further, it is important to consider the variety of topics and issues that can be addressed through neurotheology, ranging from practical to esoteric. Thus, topics in neurotheology can include those related to health and well-being; the impact of various practices (e.g. meditation or prayer) on the brain and body; the nature of consciousness; and ultimately questions pertaining to metaphysics, philosophy, theology, and epistemology.

It might be interesting to begin by considering whether we actually find an impetus towards analyzing religious experience in scientific ways in scriptural texts like the Bible.

In the Book of Daniel (verse 12–15) we can read:

> Please test your servants for ten days, and let us be given some vegetables to eat and water to drink. Then let our appearance be observed in your presence and the appearance of the youths who are eating the king's choice food; and deal with your servants according to what you see. So he listened to them in this matter and tested them for ten days. At the end of ten days their appearance seemed better and they were fatter than all the youths who had been eating the king's choice food.
>
> **(New King James Bible)**

From ancient religious texts, there is an inclination to assess the impact of religiousness on individuals, possibly even representing early attempts at controlled trials – the religious group versus the youths eating the king's choice food. While biomedical research

DOI: 10.1201/9781032717791-11

has progressed immensely since then, investigating religious phenomena remains challenging.

This chapter focuses on physiological and neurobiological investigations of religious and spiritual phenomena. While clinical work offers insights into real-life implications, physiological studies are vital for comprehending their mechanisms and understanding spirituality's impact on the body.

This chapter critically examines four dimensions pertinent to the neuroscientific exploration of religious and spiritual phenomena: 1) defining appropriate measures, 2) selecting subjects and comparison groups, 3) designing studies and applying biostatistics, and 4) addressing neurotheological and epistemological concerns.

11.2 Definition and Measurement of Spirituality and Religiousness

One key challenge in measuring religious and spiritual phenomena lies in linking subjective experiences with objective measures. For instance, if a specific meditation practice lowers blood pressure or affects brain metabolism, understanding the individual's actual experience becomes crucial.

11.2.1 Subjective Assessments

Fundamentally, the most crucial evaluations of religious and spiritual experiences involve their subjective dimensions. When individuals undergo such experiences, they often describe them through cognitive, behavioral, and emotional aspects. In spite of these various psychological and cognitive elements, people ultimately label these experiences "spiritual," distinguishing them from what they consider "non-spiritual."

Measuring this subjective aspect is like deciphering an enigmatic "black box" where events occur beyond immediate observation by outsiders. Comparing experiences across individuals and cultures presents added complexity. A spiritual experience for a Jew might differ significantly from one for a Hindu. Moreover, experiences likely span a spectrum from subtle (e.g. saying a simple prayer) to profoundly mystical experiences (d'Aquili and Newberg, 1993; Yaden and Newberg, 2023). Researchers grapple with how to quantify these subjective aspects of spiritual experiences. Developing accurate scales to measure spirituality and religiosity proves challenging, especially when relevant scales are published in less recognized, non-scientific journals (Larson, Swyers, and McCullough, 1998).

Numerous efforts have been made to create self-report scales capturing the subjective essence of specific religious or spiritual phenomena. The book *Measures of Religiosity* (Hill and Hood, 1999) offers a range of scales and questionnaires assessing aspects from commitment and awe to hope and direct experiences of the Divine. Some have undergone validity and reliability testing, essential for their utility in research (Patten, 2000).

Validity ensures that a scale measures what it claims; for instance, a valid hope scale should address hope, not just general positive feelings. Reliability determines if the scale yields consistent results over time. However, defining terms like "spiritual" or "religious" poses challenges. Different interpretations, such as spirituality referring to "awe" for one person and "oneness" to another, complicate scale development. Without clear definitions, the potential for valid scale development remains challenging. Moreover, the variability of spiritual and religious experiences within individuals further complicates reliability assessment, making it a persistent challenge in measuring spirituality accurately.

A challenge with many scales lies in their applicability across diverse traditions and cultures. Many scales cited in *Measures of Religiosity* are rooted in Christianity, potentially limiting their relevance for other traditions. Fortunately, some scales possess universal characteristics or can be adapted for broader perspectives. However, this adaptation can raise concerns about the validity and reliability of the scales in different cultural contexts.

Another intriguing issue with scales aiming to gauge the subjective nature of spiritual or religious experiences stems from their focus on psychological, affective, or cognitive processes. Questions often probe feelings, sensory experiences, or thoughts about an experience. While such measures are valuable for exploring the neural correlates of these experiences, linking psychological, affective, and cognitive elements to specific brain functions, they may fail to capture the "true" essence of spirituality.

One might argue that to access the genuine spiritual dimension, researchers should eliminate all other factors – cognitive, emotional, sensory – through analysis, isolating the purely spiritual components of the experience. For instance, an intriguing finding from a brain scan during prayer could be the absence of unique brain activity at the peak of a profound spiritual experience.

As previously mentioned, developing accurate measures hinges on ensuring they genuinely capture what they purport to measure. A subjective scale intended to assess an individual's religiosity must focus on defining religiousness and spirituality clearly. These definitions must be operationalized for effective measurement (Koenig, 1998; Koenig, McCullough, and Larson, 2001). It is crucial to avoid overly restrictive definitions that hinder research or overly broad ones that encompass unrelated experiences, such as cults or devil worship (Newberg, 2018). For instance, a definition of religiousness centered on the concept of a single God would exclude nearly two billion Hindu and Buddhist practitioners.

A consensus conference of scientists exploring spirituality and health proposed criteria for defining religiousness and spirituality (Larson, Swyers, and McCullough, 1998). The criteria for spirituality encompassed the "subjective feelings, thoughts, experiences, and behaviors arising from a search or quest for the sacred." In this criteria, "search" indicated efforts to recognize, express, maintain, or transform, while "sacred" denoted an individual's perception of a divine entity, ultimate reality, or truth. The criteria for religiousness incorporated the spirituality criteria and/or a pursuit of non-sacred objectives, like identity, belonging, meaning, health, or wellness, typically within a spiritual context. Additionally, religiousness involved having a means and methods of this search that is endorsed by a recognizable group. Thus, there are a group of people who identify what it means to be Catholic or Buddhist. While these definitions offer an operationalized approach, they may omit certain religiousness and spirituality aspects and could pose challenges from theological or philosophical viewpoints. Thus, definitions of religion and spirituality should remain adaptable to evolving findings and analyses.

Another challenge linked to definition issues is the diverse approaches to religious and spiritual phenomena, making generalization across studies difficult. Some scholars highlight differences among meditation practices or experiences (Andresen, 2000; Andresen and Forman, 2000). It is essential for studies to specify the exact practices, sub-practices, and traditions examined. Moreover, brain changes associated with one form of meditation might not apply to another. The evolving nature of this research might introduce new categorization or taxonomic methods, shedding light on whether different meditation types are genuinely distinct or merely perceived as such (Nash and Newberg, 2023).

11.3 Objective Measures of Spirituality

Objective measures of religious and spiritual experiences within neuroscience encompass a range of physiological and neurophysiological indicators. Advances in fields like psychoneuroendocrinology and psychoneuroimmunology highlight the intricate link between the brain and the body. Perceptions or feelings in the brain can impact bodily functions. While this can add complexity and introduce confounding factors, this multidisciplinary and holistic approach offers a comprehensive analysis of religious and spiritual phenomena (Newberg and Iversen, 2003). Various measures reported in research focus on autonomic nervous system (ANS) activity, commonly used to study practices like meditation or prayer. Studies have identified changes in blood pressure and heart rate during these practices (Sudsuang, Chentanez, and Veluvan, 1991; Jevning, Wallace, and Beidebach, 1992; Koenig, McCullough, and Larson, 2001). Interestingly, these changes can involve both relaxation (parasympathetic) and arousal (sympathetic) responses. Gellhorn and Kiely's early work (1972) proposed a model of meditation's physiological processes, emphasizing the importance of ANS activity. They suggested intense stimulation of either system might be part of various spiritual practices and experiences.

Early research has shown predominant parasympathetic activity during meditation, linked to decreased heart rate, blood pressure, respiratory rate, and oxygen metabolism (Sudsuang, Chentanez, and Veluvan, 1991; Jevning, Wallace, and Beidebach, 1992; Travis, 2001). However, studies on different meditation techniques indicated simultaneous activation of the parasympathetic and sympathetic systems, reflected in heart rate variability during meditation (Peng et al., 1999). This aligns with meditative states characterized by both calmness and alertness (Hugdahl, 1996; Newberg, 2018; Yaden and Newberg, 2023).

Beyond the autonomic nervous system, other studies have focused on hormone and immune function measures, particularly as complementary indicators for clinical outcomes (O'Halloran et al., 1985; Walton et al., 1995; Tooley et al., 2000; Infante et al., 2001). For instance, if meditation practices are shown to reduce cancer rates, assessing individuals' immunological and hormonal status could reveal a physiological basis. Specific cancers relate to immune (e.g., leukemia or lymphoma) or hormonal function (e.g., breast and prostate cancer). Changes in these functions may also correlate with alterations in brain processes, and this interaction can be bidirectional. Hormonal states can impact brain function, as seen in women with premenstrual syndrome, and various neurohormones can influence emotional, cognitive, and behavioral states.

Neurophysiological alterations linked to religious and spiritual experiences can be captured using various techniques, each with its pros and cons. Generally, the primary criterion is that the methodology assesses functional brain changes. There are multiple methods to measure such changes. Early research on meditation extensively employed electroencephalography (EEG) to track brain electrical activity (Banquet, 1973; Hirai, 1974; Hebert and Lehmann, 1977; Corby et al., 1978). EEG offers the advantages of being relatively non-invasive with excellent temporal resolution, allowing immediate recognition of an individual's state changes (Lehmann et al., 2001; Aftanas and Golocheikine, 2002; Travis and Arenander, 2004). However, EEG's spatial resolution is limited, showing changes over broad brain regions. Analyzing the extensive recordings from EEG sessions can also be challenging. Nonetheless, EEG can be valuable when combined with functional imaging methods, potentially indicating specific states or ensuring the subject remains awake.

Functional neuroimaging studies examining religious and spiritual phenomena have employed positron emission tomography (PET), single photon emission computed tomography (SPECT), and functional magnetic resonance imaging (fMRI) (Newberg, 2010, 2018). These techniques can detect functional brain changes in pathological conditions (e.g. neurological or psychological disorders), in response to pharmacological treatments, and during different activation states. Activation states can encompass sensory stimulation (e.g., vision, auditory), motor function and coordination, language, and higher cognitive functions (e.g., concentration) (Alavi and Newberg, 2018). These methods can measure various changes, including general physiological processes like cerebral blood flow and metabolism, as well as aspects of neurotransmitter systems. Systems such as serotonin, dopamine, opiate, benzodiazepine, glutamate, and acetylcholine have been examined across different brain states (Newberg and Alavi, 2003; Warwick, 2004; Kennedy and Zubieta, 2004).

Functional neuroimaging studies have significantly advanced our understanding of the human brain, each with its unique strengths and weaknesses when evaluating religious and spiritual phenomena. Functional MRI primarily detects changes in cerebral blood flow or oxygen levels, which correlates with brain activity. This coupling enables observation of both increased and decreased brain activity. fMRI offers excellent spatial and temporal resolution, providing precise localization of brain activation and capturing rapid brain responses, respectively. Moreover, fMRI doesn't involve radioactive exposure. However, challenges include the necessity for scanning during the study, the scanner's noise level, and assumptions regarding the correlation between blood flow and brain activity, which may not always hold true, as seen in conditions like stroke or pharmacological interventions. Additionally, fMRI cannot assess individual neurotransmitter systems unless it is performed as magnetic resonance spectroscopy (MRS). However, MRS can only look at the concentration of various molecules and not whether they are bound to receptors.

PET and SPECT imaging also have advantages and disadvantages. They provide relatively good spatial resolution, comparable to fMRI for PET and slightly lower for SPECT. While PET and SPECT images can be aligned with anatomical MRI, obtaining this requires a separate session unless a PET-MRI scanner is used (these are not widely available due to cost and maintenance requirements). Both PET and SPECT necessitate the use of a radioactive tracer, though typically at low levels. Various functional parameters can be assessed using these tracers, including blood flow, glucose or oxygen metabolism (which more precisely reflects cerebral activity), and numerous neurotransmitter components. The capability to evaluate these neurotransmitter systems sets PET and SPECT apart. Such neurotransmitter tracers can assess both state and trait responses.

Common radioactive materials like fluorodeoxyglucose, which measures glucose metabolism, can be administered through an intravenous catheter outside the scanner, facilitating practices like meditation and prayer. Subjects can be injected in a more natural setting with more natural postures and movements and then can be scanned after the conclusion of the practice while still reflecting changes that occurred during the practice. However, the temporal resolution or PET and SPECT is generally poor, ranging from minutes to hours, or even days, depending on the tracer. Thus, studying multiple prayer states in one session using PET or SPECT is generally not possible. Nevertheless, capturing two or three states in a single imaging session is feasible with the right radiopharmaceutical (Lou et al., 1999). Ultimately, the choice of neuroimaging technique should align with the study's specific objectives.

Several overarching challenges hinder the interpretation of results from functional neuroimaging studies. A primary concern is determining precisely what physiological changes are being measured and how they correlate with subjective experiences. Interpreting specific scan findings, such as increased activity in inhibitory neurons, presents uncertainties regarding their impact on cerebral activity as measured by PET or fMRI. Another challenge lies in aligning these physiological changes with subjective states, especially when it is impossible to intervene during peak experiences to gauge feelings directly.

Additionally, the brain undergoes various changes during such experiences, each with different intensities. Determining which degree of change is considered significant remains ambiguous (e.g., a 10% or 20% change). Statistical challenges further complicate the analysis, including comparing images across subjects and conditions, addressing multiple comparison issues, and interpreting activation states and individual brain regions. Statistical parametric mapping (SPM) offers a solution by normalizing and co-registering images, then analyzing them pixel by pixel for significant changes (Friston et al., 1995). However, due to the conservative nature of SPM, subtle changes might be overlooked, raising questions about their clinical relevance.

Despite these challenges, neuroimaging studies have effectively assessed specific spiritual and meditative practices. Several studies employing different neuroimaging techniques have shown consistency in their findings, for example, highlighting the involvement of the frontal lobes, parietal lobes, thalamus, and the limbic system in these practices (Herzog, 1990–1991; Lou et al., 1999; Lazar et al., 2000; Newberg et al., 2001a, 2003; Kjaer et al., 2002; Newberg and Iversen, 2003; Yaden and Newberg, 2023). Interestingly, these studies indicate that various practices may activate a complex network of brain structures (Newberg, 2018; Yaden and Newberg, 2023). Moreover, differences observed seem to align with variations among the practices studied. For instance, one study found increased dopamine activity during meditation-related practices (Kjaer et al., 2002). As research progresses, our understanding of the neurophysiological changes associated with religious and spiritual phenomena will undoubtedly deepen, necessitating further studies for a comprehensive understanding.

11.4 The Importance of Subject Selection and Comparison Groups

Determining the most suitable subjects and comparison groups is a crucial methodological consideration when studying religious and spiritual phenomena. The choice of subjects largely hinges on the specific definition of the phenomenon under investigation. For instance, if a researcher aims to examine physiological changes during yoga meditation, the study group would ideally comprise individuals who practice this specific form of meditation. Conversely, when exploring feelings of unity, the study might involve participants from diverse religious and cultural backgrounds.

The expertise or proficiency level of the participants is another vital factor to consider. Variations in results might arise between novice, experienced, and master-level practitioners due to different physiological processes associated with each level. Additionally, environmental factors, such as the noise in an MRI scanner, could affect performance consistency among participants. For example, novices might find it more challenging to maintain their meditation practice under such distractions compared to experienced practitioners. Therefore, observed differences may not necessarily reflect inherent distinctions between practices but rather variations in their execution under research conditions.

Assessing socioeconomic and health backgrounds, age, gender, sexual orientation, and other lifestyle factors among participants is also crucial. Differences in lifestyle habits, such as smoking, could introduce confounding variables, potentially influencing brain scan results.

Regarding comparison or control groups, some studies use the participants as their own comparison, contrasting the meditation state with their baseline waking state.

Alternatively, comparisons can be made against tasks that are similar but lack specific spiritual significance. For instance, a study comparing the effects of different mantras on brain electrical activity during meditation found distinct differences between spiritual and non-spiritual mantras (Telles et al., 1998).

When participants serve as their own controls, it is essential to account for all practice-related factors. For instance, if a spiritual practice involves burning incense or reading prayers with open eyes, the baseline state should replicate these conditions. Failing to do so might lead to misinterpretations, such as attributing changes in olfactory regions to the spiritual practice when they are due to the incense's aroma. Similarly, studies have found differences in cerebral activity based on whether participants read religious or non-religious texts (Azari et al., 2001).

Other potential comparison groups might include practitioners from the same tradition with varying expertise levels or practitioners from different traditions practicing similar activities. While these groups could provide insights into the long-term effects of spiritual practices, factors like age and health could complicate the interpretation of study results.

Placebo groups pose a significant challenge in studies focusing on religious and spiritual phenomena. Determining an appropriate "placebo" can be complex, especially since individuals with spiritual beliefs usually recognize whether they are engaging in genuine spiritual practices. In such cases, placebo groups often involve tasks that mimic the spiritual activity but lack its specific spiritual content. For example, if the study focuses on singing prayers, a suitable comparison might involve singing non-religious songs.

Another fascinating possibility is whether a study can actually measure God as a subject. Of course, any such study would be extremely challenging and likely result in more questions than answers.

Currently, this remains a matter of interpretation, and many scholars have attempted to use such studies either to validate or invalidate religious beliefs. Science presently appears incapable of definitively addressing supernatural claims. For instance, in studies examining intercessory prayer, successful outcomes could be attributed to divine intervention or the potential influence of human consciousness on remote individuals. Conversely, a lack of significant results in such studies does not necessarily disprove the existence of God.

11.5 Considerations of Study Design and Analysis

Drawing from the existing literature and the proposed definition of spiritual experience, several neuroscientific paradigms can be applied to operationalize spiritual experiences effectively (Larson, Swyers, and McCullough, 1998). These paradigms include but are not limited to: 1) the neurophysiology of spiritual interventions;

2) spiritual interventions linked with psychopharmacological agents; 3) drug-induced spiritual experiences; 4) neuropathologic and psychopathologic spiritual experiences; 5) spiritual experiential development in infants, children, and adolescents; and 6) physical and psychological therapeutic interventions. Following a consideration of these study designs, the biostatistical challenges associated with these studies can be examined.

11.5.1 The Neurophysiology of Spiritual Interventions

This paradigm focuses on directly studying spiritual interventions, such as prayer or meditation, alongside concurrent assessments of psychological and spiritual outcomes. This approach aims to characterize the nature of the spiritual intervention. Subsequently, these psychological and spiritual metrics can be juxtaposed with neurobiological indicators, including electroencephalographic activity, cerebral blood flow, cerebral metabolism, and neurotransmitter activity. State-of-the-art imaging techniques like PET, SPECT, and fMRI can be employed for these measurements. Physiological parameters, such as blood pressure, body temperature, heart rate, and galvanic skin responses reflecting autonomic nervous system activity, can also be monitored. Additionally, comprehensive analysis should encompass immunological assessments and hormonal levels to fully understand the impacts of spiritual interventions.

11.5.2 Altering Spiritual Interventions

Another paradigm for investigating spiritual experiences neuroscientifically involves the use of pharmacological agents or interventions to modify spiritual practices. In this paradigm, a spiritual intervention is compared before and after the introduction of an additional intervention. For instance, researchers might explore the impact of an opiate antagonist on the intensity of meditation or prayer experiences. Initial studies on this topic showed no changes in EEG patterns during meditation when subjects received opiate or benzodiazepine antagonists (Sim and Tsoi, 1992). Further studies might assess the effects of transcranial magnetic stimulation, various medications, or even surgical procedures (performed for other reasons). However, comprehensive studies involving multiple neurophysiological parameters are necessary. Researchers may employ different agonists and antagonists to examine their potential to enhance or reduce spiritual experiences. Additionally, investigating various pharmacological agents could clarify the roles of different neurotransmitter systems, providing insights into dose-response relationships in spiritual interventions.

11.5.3 Drug-Induced Spiritual Experiences

A third paradigm considers individuals who have experienced intense spiritual episodes due to hallucinogenic drug use. Observations suggest that drugs like opiates, LSD, and stimulants can induce spiritual experiences. Detailed studies of drug-induced spiritual experiences, possibly using modern imaging techniques, could shed light on the neurobiological mechanisms underpinning naturally occurring spiritual experiences. Some studies on this topic have been conducted (Vollenweider et al., 1997, 1999, 2000; Yaden et al., 2016), but more extensive research is needed, especially concerning religious and spiritual experiences. Contrasting psychedelic with naturally occurring spiritual experiences could help distinguish pathological from non-pathological spiritual experiences. While ethical and legal considerations are significant, much interest in the effect of psychedelic compounds in the management of psychiatric problems is ongoing. Subjects who have experienced drug-induced spiritual episodes could be studied using radioactive analogs of these agents to determine receptor concentrations and their agonists. Alternatively, fMRI or EEG studies while subjects are under the influence of psychedelic compounds can be performed. Another avenue could be exploring the effects of drug withdrawal on spiritual experiences, although such findings are currently absent in the literature.

11.5.4 Neuropathological- and Psychopathological-Related Spiritual Experiences

A fourth approach involves studying patients with established neuropathologic and psychopathologic conditions. Neurological conditions, including seizure disorders, especially in the temporal lobes, brain tumors, and strokes, have been linked to spiritual experiences or shifts in religious beliefs. Temporal lobe epilepsy, for instance, has been linked to increased religiosity and religious conversions (Bear, 1979; Bear and Fedio, 1977). Similarly, psychiatric conditions like schizophrenia and mania have also been associated with spiritual experiences and religious conversions. Identifying the specific pathology and its location can help uncover the neurobiological basis of spiritual experiences. Thus, neuropsychiatric disorders offer valuable insights into the neuroscience of spiritual experiences.

Historically, research on pathological conditions has helped clarify the normal functions of biological systems. Spiritual experiences in psychiatric and neurological disorders could be key to identifying emerging neurobiological systems that support "normal" spiritual experiences. This challenges the traditional psychiatric notion suggesting that spiritual experiences reflect psycho- or neuropathology. Such an approach allows for the possibility of both normal and pathological spiritual experiences in individuals with or without psychopathological conditions. However, caution is needed to avoid exclusively associating spiritual experiences with pathology or reducing them solely to neurophysiological mechanisms.

11.5.5 Spiritual Experiential Development

There is substantial literature on the developmental aspects of religion and spiritual experience, spanning infancy to adulthood (Fowler, 1981; Tamminen, 1994; Newberg and Newberg, 2005). These studies explore the progression of spiritual experiences and their neurocognitive foundations. For instance, rudimentary, undifferentiated faith might manifest in infancy, while the multifaceted nature of spiritual experience, encompassing cognitive, cultural, and emotional components, often develops fully in adulthood (Fowler, 1981). Most analyses of spiritual development are grounded in psychology, but neuroscience could leverage these insights by comparing them with the development of various brain structures and neurocognitive processes. This approach may help identify the brain structures and functions necessary for different aspects of spiritual experience. Furthermore, examining the end-of-life perspective could shed light on changes in spiritual functions linked to diffuse neuropathological conditions like dementia. Studying alterations in spiritual functions related to declines in neurocognitive and physical health could also be beneficial.

11.5.6 Physical and Psychological Therapeutic Interventions

Numerous studies have investigated the therapeutic benefits of practices like meditation, stress management, and prayer for a range of psychological and physical conditions, including anxiety disorders, hypertension, coronary artery disease, cancer, and HIV (Kabat-Zinn et al., 1992; Carson, 1993; Levin, 1994; Levin and Vanderpool, 1989; Miller et al., 1995; Leserman et al., 1989; Zamarra et al., 1996; Massion et al., 1995; Schneider et al., 1995). While these studies primarily focus on the intervention's impact on disease parameters, there's potential to incorporate measures of spiritual experience and well-being. Using existing measurement scales, researchers might explore the relationship between spiritual experiences, well-being, and the intervention's effect on disease progression. High-quality research is crucial for establishing the link between spirituality and health, addressing methodological concerns raised in earlier studies (Sloan, Bagiella, and Powell, 1999; Sloan and Bagiella, 2002).

11.5.7 Statistical Analysis Issues

Statistical analysis in the study of religious and spiritual phenomena presents unique challenges. These phenomena are often multifaceted, encompassing both subjective and objective components. Given this complexity, statistical analyses need to consider numerous variables, ensuring that findings are not oversimplified and that essential covariates are not overlooked. It is crucial to approach statistical analyses of religious and spiritual phenomena with the same rigor and complexity as biomedical studies, employing qualified statisticians to uphold research quality.

Additionally, religious phenomena may involve distinct interindividual differences worth examining. For instance, in our prior study of Tibetan Buddhist meditators, participants practiced the same meditation type for identical durations to facilitate group data pooling. Yet this approach might have overlooked individual variations in meditation depth, experiences, and unique techniques employed. Statistical analyses can struggle to assess these interindividual differences, especially concerning physiological measures. Developing analytical methods to better explore these differences could significantly advance the field.

11.6 Neuroscientific Models of Spiritual Experiences

When it comes to neuroscientific models of religious and spiritual experiences, it might be best to consider a larger network of structures and functions that appear to be involved. Early models focused on major brain structures, such as the frontal, temporal, and parietal lobes. However, as research has progressed, it has become clear that there is a complex network of structures that become activated in various patterns, depending on the practices and experiences involved.

The basic components of such models is provided subsequently. it is important to emphasize that most spiritual practices are dynamic in that there is a progression of changes and experiences that occurs overtime. The experiences during the first minutes of meditation are substantially different compared to those that arise one hour, one day, or one month after beginning an intensive meditation program.

Frontal lobe activation has been a prominent finding in many practices. The frontal lobes are typically involved in executive processes and are part of a larger attentional network. Since many spiritual practices, such as meditation or prayer, involve the intense focus of attention on a given object, such as a mantra or prayer, frontal lobe activity has frequently been observed on brain scan studies (Newberg et al., 2001a, 2001b; Newberg and Iversen, 2003).

The parietal lobes have also been implicated as having an important role when it comes to spatial and temporal changes that are experienced during spiritual practices. Research has suggested that decreased parietal activity occurs during many different spiritual practices. Since parietal lobe function is involved in integrating sensory information to produce a spatial representation of the self, decreased parietal lobe activity is likely associated with a loss of the sense of self. This is frequently described during spiritual practices, such as meditation and prayer (Yaden and Newberg, 2023). In addition to the loss of the sense of self, a sense of connectedness and a sense of oneness are frequently described during these experiences. Neuroimaging studies have revealed decreased parietal lobe function during a variety of spiritual practices. In addition, lesion studies have found that those individuals with injury to the parietal lobes are more likely to express feelings of self-transcendence compared to lesions that occur in other parts of the brain (Urgesi et al., 2010).

The cortical areas of the temporal lobe are involved in language as well as visual processing. Practices such as prayer that involve the repetition of specific verbal passages, or the engagement in a "conversation with God," may all be associated with increased activity in the superior temporal regions. Similarly, the inferior temporal lobe is involved in visual processing and may be associated with practices that involve focus on religious imagery, such as the cross or other symbols.

The medial portion of the temporal lobe incorporates the insula and limbic system, which are prominently involved in emotional responses. Many spiritual and religious practices and experiences include profound, emotional experiences, such as joy, awe, love, or fear. A number of studies have demonstrated increased activity in the limbic structures during intense, spiritual experiences (Newberg, 2018). Changes in the brainstem and autonomic nervous system have been described, and also referred to previously. Intense feelings of arousal or calmness are likely to be associated with physiological changes in these areas of the nervous system (Newberg, 2018).

Many of these structures are also considered to be part of brain networks, such as the salience network, default mode network, and attentional network. A number of studies have revealed components of these networks to be involved in spiritual practices (Brewer et al., 2011; Guu et al., 2023; Zhang et al., 2023).

In addition to the brain regions described, there are a number of neurotransmitter changes that have been observed during meditation and prayer practices. Neurotransmitters such as dopamine and serotonin are involved in emotional and cognitive processing. The dopamine regions are involved in the reward system of the brain that elicit positive responses and reinforce beliefs and behaviors. Thus, the release of dopamine may be an important part of enhancing religious experiences. Serotonin is the target of many psychedelic drugs and appears to be associated with intense mystical or spiritual experiences. While sensory elements are prominent with psychedelic compounds, many of the other aspects of spiritual experiences may also be described. Other excitatory and inhibitory neurotransmitters such as glutamate and gamma aminobutyric acid have also been described to play a role (Yaden et al., 2017; Newberg, 2018).

In the end, it is important to recognize that religious and spiritual practices and experiences do not involve a single part of the brain, but a complex network involving many structures and physiological processes. This would be expected given the richness and diversity of such experiences across traditions and people.

11.7 Neurotheological and Epistemological Implications

A longstanding philosophical quandary that lies at the heart of neurotheology asks, "How can we be sure the external world aligns, at least somewhat, with our mental perceptions, particularly when it comes to religion and spirituality?" This question about what constitutes "true reality" dates back to ancient Greek philosophers and is even more ancient in Eastern religious and philosophical traditions. When considering the neuroscientific study of religious and spiritual phenomena, it is worth questioning if epistemological concerns can truly be resolved.

Some researchers argue that identifying a neurobiological correlate for a religious experience dismisses its significance. However, merely identifying such a correlate does not explain the underlying cause of these phenomena. For instance, changes in brain activity during a mystical experience do not clarify if the brain instigated or responded to that event. Similarly, inducing religious states with pharmacological agents does not diminish their spiritual essence. Shamanic practices involving substance use for spiritual journeys are not considered less authentic due to these substances. Interestingly, substances alone can be associated with profound spiritual experiences, although it is less clear whether they would also have more traditional religious meaning for the individual (Yaden et al., 2016).

Thus, the context in which these practices and experiences occur remains essential to their spiritual significance.

Revisiting the epistemological query through a neuroscientific lens introduces a concept termed "neuroepistemology," revealing a complex paradox in how the brain interprets reality (Newberg, d'Aquili, and Rause, 2001b; Newberg, 2010, 2018). The most common criteria used to determine reality include:

1. Subjective vividness of reality.

2. Duration over time.

3. Intersubjective agreement on what's real.

Each criterion can be linked to specific brain functions. However, it can be argued that all three criteria ultimately stem from the first – the vivid sense of reality. For instance, our perception of time depends on how our brain structures it. Distorting the brain's temporal ordering can profoundly alter our sense of time, evident during spiritual practices where time seems to vanish. This challenges the notion of time as an absolute, suggesting it is shaped by brain structures, a notion which is explicitly articulated in some religious traditions, such as various forms of Hinduism.

Similarly, the third criterion – intersubjective validation – also raises questions. The entities we perceive as real are essentially reflections within our cognitive framework. Therefore, anyone analyzing their own experiences may initially operate from a solipsistic standpoint. Thus, basing reality on characteristics of our perception, such as time, which is inherently structured by the brain, can lead to circular reasoning.

These discussions could be extended endlessly. However, it seems evident that every criterion for determining the reality of entities ultimately converges on the first criterion – the vivid sense of reality. This leads to intriguing questions about religious and spiritual states that feel more real than our everyday reality, even when remembered from within that baseline reality.

When viewed from our everyday perspective, some states may seem less real than baseline reality, while others appear more real when recalled. For instance, dreams and psychotic hallucinations are often perceived as less real upon reflection in a normal state. In contrast, many religious and spiritual experiences, like absolute unitary states (Newberg, d'Aquili, and Rause, 2001b), Cosmic Consciousness as described by Bucke (1961), certain trance states, hyperlucid visions, and near-death experiences (Newberg and d'Aquili, 1994), seem more vividly real when remembered within baseline reality.

These experiences can profoundly impact individuals, leading to sustained changes in behavior and attitudes, such as increased altruism and reduced fear of death.

Research on near-death experiences suggests that these transformative effects may last for years, or perhaps even a lifetime (Greyson, 2022). If reality ultimately boils down to our vivid sense of it, these states, particularly absolute unitary states or pure consciousness, could indeed be more real than our everyday baseline reality. Here, "real" is not used poetically but parallels the straightforward sense of calling a rock or table "real."

Approaching reality from a neuroscientific viewpoint reveals its elusive nature, often challenging our intuitive understanding and presenting profound insights to scientists, philosophers, and mystics alike. Such an approach is fundamental to the field of neurotheology.

Neurotheology refers to the field of study that explores this relationship between the brain and spiritual and religious phenomena. It requires scholars to be open and understanding of both the scientific and spiritual sides of this field. Science should be kept as rigorous as possible and strive to find optimal methodologies addressing many of the factors mentioned previously. Further, religious and spiritual phenomena must be seriously considered for their characteristics and also what they say about the nature of humanity and the nature of reality. As mentioned, spiritual experiences are often described as representing a more fundamental perspective of reality, one that goes beyond the objective and subjective aspects of reality. While it is not clear how much weight should be given to both the scientific and spiritual sides of neurotheology, careful consideration of all possibilities should be employed.

11.8 Conclusion

While the study of religious and spiritual phenomena through neuroscience has made significant strides since its inception over three decades ago, it remains a nascent area of research. The field grapples with unique methodological challenges, alongside broader issues related to funding and academic recognition. Nonetheless, investing in such endeavors holds the promise of substantial rewards for both scientific and religious understanding.

From a religious standpoint, these studies offer insights into the human experience of spirituality and religion. They deepen our understanding of how religious practices influence the mind, brain, body, and behavior. On the scientific front, this research provides opportunities to explore the intricate workings of the human brain and understand the relationship between brain states and bodily functions.

Addressing methodological and statistical challenges can further enrich both scientific and spiritual domains. It can lead to advancements in scientific and statistical methodologies while also fostering theological and philosophical discussions. Ultimately, the convergence of neuroscience and religion promises to be a pivotal area of scholarly exploration in the 21st century and beyond, bridging the gap between spirituality and scientific inquiry.

References

Aftanas LI, Golocheikine SA. (2002). Non-linear dynamic complexity of the human EEG during meditation. *Neurosci Lett* 330(2):143–146.

Alavi A, Newberg AB. (2018). Functional neuroimaging: A transformative tool for integrative psychiatry. In Monti DA, Newberg AB (editors). *Integrative Psychiatry and Brain Health*. New York, NY: Oxford University Press, pp. 347–381.

Andresen J. (2000). Meditation meets behavioural medicine – the story of experimental research on meditation. *J Conscious Stud* 7(11–12):17–73.

Andresen J, Forman RKC. (2000). Methodological pluralism in the study of religion – how the study of consciousness and mapping spiritual experiences can reshape religious methodology. *J Conscious Stud* 7(11–12):7–14.

Azari NP, Nickel J, Wunderlich G, Niedeggen M, Hefter H, Tellmann L, Herzog H, Stoerig P, Birnbacher D, Seitz RJ. (2001). Neural correlates of religious experience. *Eur J Neurosci* 13(8):1649–1652.

Banquet JP. (1973). Spectral analysis of the EEG in meditation. *Electroencephalogr Clin Neurophysiol* 35(2):143–151.

Bear DM. (1979). Temporal lobe epilepsy – a syndrome of sensory-limbic hyperconnection. *Cortex* 15:357–384.

Bear DM, Fedio P. (1977). Quantitative analysis of interictal behavior in temporal lobe epilepsy. *Arch Neurol* 34:454–467.

Brewer JA, Worhunsky PD, Gray JR, Tang YY, Weber J, Kober H. (2011). Meditation experience is associated with differences in default mode network activity and connectivity. *Proc Natl Acad Sci USA*. 108(50):20254–20259.

Bucke RM. (1961). *Cosmic Consciousness*. Secaucus, NJ: Citadel Press.

Carson VB. (1993). Prayer, meditation, exercise, and special diets: Behaviors of the hardy person with HIV/AIDS. *JANAC* 4:18–28.

Corby JC, Roth WT, Zarcone VP Jr, Kopell BS. (1978). Psychophysiological correlates of the practice of tantric yoga meditation. *Arch Gen Psychiatry* 35:571–577.

d'Aquili EG, Newberg AB. (1993). Religious and mystical states: A neuropsychological model. *Zygon* 28:177–199.

Fowler JW. (1981). *Stages of Faith*. San Francisco: Harper Collins Publishers.

Friston KJ, Holmes AP, Worsley KJ, Poline JP, Frith CD, Frackowiak RSJ. (1995). Statistical parametric maps in functional imaging: A general linear approach. *Human Brain Mapp* 2:189–210.

Gellhorn E, Kiely WF. (1972). Mystical states of consciousness: Neurophysiological and clinical aspects. *J Nerv Mental Dis* 154:399–405.

Greyson B. (2022). *After: A Doctor Explores What Near Death Experiences Reveal About Life and Beyond*. New York: St. Martin's Essentials.

Guu SF, Chao YP, Huang FY, et al. (2023). Interoceptive awareness: MBSR training alters information processing of salience network. *Front Behav Neurosci* 17:1008086. https://doi.org/10.3389/fnbeh.2023.1008086.

Hebert R, Lehmann D. (1977). Theta bursts: An EEG pattern in normal subjects practising the transcendental meditation technique. *Electroencephalogr Clin Neurophysiol* 42(3):397–405.

Herzog H., Lele V.R., Kuwert T, et al. (1990–1991). Changed pattern of regional glucose metabolism during Yoga meditative relaxation. *Neuropsychobiology* 23: 182–187.

Hill PC, Hood RW. (1999). *Measures of Religiosity*. Religious Education Press.

Hirai T. (1974). *Psychophysiology of Zen*. Tokyo: Igaku Shoin.

Hugdahl K. (1996) Cognitive influences on human autonomic nervous system function. *Curr Opin Neurobiol* 6:252–258.

Infante JR, Torres-Avisbal M, Pinel P, Vallejo JA, Peran F, Gonzalez F, Contreras P, Pacheco C, Roldan A, Latre JM. (2001). Catecholamine levels in practitioners of the transcendental meditation technique. *Physiol Behav* 72(1–2):141–146.

Jevning R, Wallace RK, Beidebach M. (1992). The physiology of meditation: A review. A wakeful hypometabolic integrated response. *Neurosci Biobehav Rev* 16:415–424.

Kabat-Zinn J, Massion AO, Kristeller J, Peterson LG, Fletcher KE, Pbert L, Lenderking WR, Santorelli SF. (1992). Effectiveness of a meditation-based stress reduction program in the treatment of anxiety disorders. *Am J Psychiatry* 149:936–943.

Kennedy SE, Zubieta JK. (2004). Neuroreceptor imaging of stress and mood disorders. *CNS Spectr* 9(4):292–301.

Kjaer TW, Bertelsen C, Piccini P, Brooks D, Alving J, Lou HC. (2002). Increased dopamine tone during meditation-induced change of consciousness. *Brain Res Cogn* 13(2):255–259.

Koenig HG. (editor). (1998). *Handbook of religion and mental health.* San Diego: Academic Press.

Koenig HG, McCullough ME, Larson DB. (editors). (2001). *Handbook of Religion and Health.* New York: Oxford University Press.

Larson DB, Swyers JP, McCullough ME. (editors). (1998). *Scientific Research on Spirituality and Health: A Consensus Report.* Washington, DC: National Institute for Healthcare Research.

Lazar SW, Bush G, Gollub RL, Fricchione GL, Khalsa G, Benson H. (2000). Functional brain mapping of the relaxation response and meditation. *Neuroreport* 15;11(7):1581–1585.

Lehmann D, Faber PL, Achermann P, Jeanmonod D, Gianotti LR, Pizzagalli D. (2001). Brain sources of EEG gamma frequency during volitionally meditation-induced, altered states of consciousness, and experience of the self. *Psychiatry Res* 108(2):111–121.

Leserman J, Stuart EM, Mamish ME, Benson H. (1989). The efficacy of the relaxation response in preparing for cardiac surgery. *Behav Med* 15:111–117.

Levin JS. (1994). Religion and health: Is there an association, is it valid, and is it causal? *Soc Sci Med* 38(11):1475–1482.

Levin JS, Vanderpool HY. (1989). Is religion therapeutically significant for hypertension? *Soc Sci Med* 29(1):69–78.

Lou HC, Kjaer TW, Friberg L, Wildschiodtz G, Holm S, Nowak M. (1999). A 15O-H2O PET study of meditation and the resting state of normal consciousness. *Hum Brain Mapp* 7(2):98–105.

Massion AO, Teas J, Hebert JR, Wertheimer MD, Kabat-Zinn J. (1995). Meditation, melatonin and breast/prostate cancer: Hypothesis and preliminary data. *Med Hypothesis* 44:39–46.

Miller JJ, Fletcher K, Kabat-Zinn J. (1995) Three-year follow-up and clinical implications of a mindfulness meditation-based stress reduction intervention in the treatment of anxiety disorders. *Gen Hosp Psychiatr* 17(3):192–200.

Nash JD, Newberg AB. (2023). An updated classification of meditation methods using principles of taxonomy and systematics. *Front Psychol* 13:1062535. https://doi.org/10.3389/fpsyg.2022.1062535.

Newberg A, Pourdehnad M, Alavi A, d'Aquili E. (2003). Cerebral blood flow during meditative prayer: Preliminary findings and methodological issues. *Percept Mot Skills* 97:625–630.

Newberg AB. (2010). *Principles of Neurotheology.* Surrey, UK: Ashgate Publishing Ltd.

Newberg AB. (2018). *Neurotheology: How Science Can Enlighten Us About Spirituality.* New York, NY: Columbia University Press.

Newberg AB, Alavi A. (2003). Role of positron emission tomography in the investigation of neuropsychiatric disorders. In Sandler MP, Coleman RE, Patton JA, Wackers FJT, Gottschalk A, Hoffer PB (editors). *Diagnostic Nuclear Medicine*, Fourth edition. Philadelphia: Lippincott Williams & Wilkins, pp. 783–819.

Newberg AB, Alavi A, Baime M, Pourdehnad M, Santanna J, d'Aquili EG. (2001a). The measurement of regional cerebral blood flow during the complex cognitive task of meditation: A preliminary SPECT study. *Psychiatry Res Neuroimaging* 106:113–122.

Newberg AB, d'Aquili EG, Rause VP. (2001b). *Why God Won't Go Away: Brain Science and the Biology of Belief*. New York: Ballantine Publishing Group.

Newberg AB, d'Aquili EG. (1994). The near death experience as archetype: A model for 'prepared' neurocognitive processes. *Anthropol Conscious* 5:1–15.

Newberg AB, Iversen J. (2003). The neural basis of the complex mental task of meditation: Neurotransmitter and neurochemical considerations. *Med Hypothesis* 61(2):282–291.

Newberg AB, Newberg SK. (2005). A neuropsychological perspective on spiritual development. In Benson PL, Roehlkepartain EC, King PE, Wagener LM (editors). *Handbook of Spiritual Development in Childhood and Adolescence*. London: Sage Publications, Inc., pp. 183–196.

O'Halloran JP, Jevning R, Wilson AF, Skowsky R, Walsh RN, Alexander C. (1985). Hormonal control in a state of decreased activation: Potentiation of arginine vasopressin secretion. *Physiol Behav* 35(4):591–595.

Patten MD. (2000). *Understanding Research Methods*, Second edition. Los Angeles: Pyrczak Publishing.

Peng CK, Mietus JE, Liu Y, et al. (1999). Exaggerated heart rate oscillations during two meditation techniques. *Intern J Cardiol* 70:101–107.

Schneider RH, Staggers F, Alexander CN, et al. (1995). A randomized controlled trial of stress reduction for hypertension in older African Americans. *Hypertension* 26:820–827.

Sim MK, Tsoi WF. (1992). The effects of centrally acting drugs on the EEG correlates of meditation. *Biofeedback and Self-Regulation* 17:215–220.

Sloan RP, Bagiella E. (2002). Claims about religious involvement and health outcomes. *Ann Behav Med* 24(1):14–21.

Sloan RP, Bagiella E, Powell T. (1999). Religion, spirituality, and medicine. *Lancet*. 353(9153):664–667.

Sudsuang R, Chentanez V, Veluvan K. (1991). Effects of Buddhist meditation on serum cortisol and total protein levels, blood pressure, pulse rate, lung volume and reaction time. *Physiol Behav* 50:543–548.

Tamminen K. (1994). Religious experiences in childhood and adolescence: A viewpoint of religious development between the ages of 7 and 20. *Int J Psychol Relig* 4:61–85.

Telles S, Nagarathna R, Nagendra HR. (1998, Jan). Autonomic changes while mentally repeating two syllables – one meaningful and the other neutral. *Indian J Physiol Pharmacol* 42(1):57–63.

Tooley GA, Armstrong SM, Norman TR, Sali A. (2000). Acute increases in night-time plasma melatonin levels following a period of meditation. *Biol Psychol* 53(1):69–78.

Travis F. (2001). Autonomic and EEG patterns distinguish transcending from other experiences during Transcendental Meditation practice. *Int J Psychophysiol* 42:1–9.

Travis F, Arenander A. (2004, Jan–Feb). EEG asymmetry and mindfulness meditation. *Psychosom Med* 66(1):147–148.

Urgesi C, Aglioti SM, Skrap M, Fabbro F. (2010). The spiritual brain: Selective cortical lesions modulate human self-transcendence. *Neuron* 65(3):309–319.

Vollenweider FX, Leenders KL, Scharfetter C, Maguire P, Stadelmann O, Angst J. (1997, May). Positron emission tomography and fluorodeoxyglucose studies of metabolic hyperfrontality and psychopathology in the psilocybin model of psychosis. *Neuropsychopharmacology* 16(5): 357–372.

Vollenweider FX, Vontobel P, Hell D, Leenders KL. (1999, May). 5-HT modulation of dopamine release in basal ganglia in psilocybin-induced psychosis in man – a PET study with [11C] raclopride. *Neuropsychopharmacology* 20(5):424–433.

Vollenweider FX, Vontobel P, Oye I, Hell D, Leenders KL. (2000, Jan–Feb). Effects of (S)-ketamine on striatal dopamine: A [11C]raclopride PET study of a model psychosis in humans. *J Psychiatr Res* 34(1):35–43.

Walton KG, Pugh ND, Gelderloos P, Macrae P. (1995). Stress reduction and preventing hypertension: Preliminary support for a psychoneuroendocrine mechanism. *J Altern Complement Med* 1(3):263–283.

Warwick JM. (2004). Imaging of brain function using SPECT. *Metab Brain Dis* 19(1–2):113–123.

Yaden DB, Iwry J, Newberg AB. (2017). Neuroscience and religion: Surveying the field. In Clements NK (editor). *Religion: Mental Religion*. Farmington Hills, MI: Macmillan Reference USA, pp. 277–299.

Yaden DB, Newberg AB. (2023). *The Varieties of Spiritual Experience*. New York: Oxford University Press.

Yaden DB, Nguyen KDL, Kern ML, Belser AB, Eichstaedt JC, Iwry J, Wintering N, Hood RW, Newberg AB. (2016). Of roots and fruits: A comparison of psychedelic and non-psychedelic mystical experiences. *J Humanist Psychol* 57(4):338–353.

Zamarra JW, Schneider RH, Besseghini I, Robinson DK, Salerno JW. (1996). Usefulness of the transcendental meditation program in the treatment of patients with coronary artery disease. *Amer J Cardiology* 77:867–870.

Zhang Y, Chen S, Zhang Z, et al. (2023). Effect of meditation on brain activity during an attention task: A comparison study of ASL and BOLD task fMRI. *Brain Sci* 13(12):1653. https://doi.org/10.3390/brainsci13121653

Index

Note: Page numbers in *italics* indicate a figure and page numbers in **bold** indicate a table on the corresponding page. Page numbers followed by "n" with numbers refer to notes.

For Product Safety Concerns and Information please contact our EU
representative GPSR@taylorandfrancis.com
Taylor & Francis Verlag GmbH, Kaufingerstraße 24, 80331 München, Germany

www.ingramcontent.com/pod-product-compliance
Ingram Content Group UK Ltd.
Pitfield, Milton Keynes, MK11 3LW, UK
UKHW052221190926
475702UK00006B/1144